POWER SYSTEM STABILITY
Analysis by the Direct Method of Lyapunov

NORTH-HOLLAND SYSTEMS AND CONTROL SERIES
VOLUME 3

NORTH-HOLLAND PUBLISHING COMPANY
AMSTERDAM • NEW YORK • OXFORD

POWER SYSTEM STABILITY

Analysis by the Direct Method of Lyapunov

M. A. PAI

Professor
Indian Institute of Technology
Kanpur, India

Visiting Professor
Iowa State University
Ames, Iowa, U.S.A.

NORTH-HOLLAND PUBLISHING COMPANY
AMSTERDAM • NEW YORK • OXFORD

ISBN: 0 444 86310 9

Publishers:

NORTH-HOLLAND PUBLISHING COMPANY
AMSTERDAM • NEW YORK • OXFORD

Sole distributors for the U.S.A. and Canada:

ELSEVIER NORTH-HOLLAND, INC.
52 VANDERBILT AVENUE
NEW YORK, N.Y. 10017

Library of Congress Cataloging in Publication Data

Pai, M. A., 1931-
Power system stability.

(North-Holland systems and control series ; v. 3)
1. Electric power system stability. 2. Liapunov functions. I. Title. II. Series.
TK1005.P33 621.319'13 81-16802
ISBN 0-444-86310-9 AACR2

PRINTED IN THE NETHERLANDS

Dedicated to
Indian Institute of Technology,
Kanpur, India

PREFACE

Research work in applying Lyapunov Stability theory to power systems has been pursued actively for the last decade and a half. The amount of literature is quite vast and there is now strong evidence that the method may find eventual application in planning and on-line dynamic security assessment. It was, therefore, felt that it is now appropriate to collate, unify and present in a cohesive conceptual framework both the theory and application aspects of the direct method of stability analysis for power systems.

After a physical statement of the problem in the introductory chapter, Lyapunov's stability theory is reviewed in Chapter Two. We develop the mathematical models for multi-machine power systems in Chapter Three stressing both the center of angle formulation and the Luré formulation of the equations. The systematic construction of Lyapunov functions for single and multi-machine power systems is the theme of Chapter Four. It is shown that the energy function is one of the many Lyapunov functions that can be constructed. The computation of the region of attraction of the post-fault stable equilibrium point is the most difficult problem in Lyapunov stability analysis of power systems. The recent approaches which make use of the faulted dynamics of the system are emphasized in Chapter Five as opposed to earlier results which gave conservative estimates of the critical clearing time. Two case studies are presented. In Chapter Six we discuss the stability of large scale power systems by decomposition (often termed as the vector Lyapunov function approach). Both the weighted sum scalar Lyapunov function and the comparison principle-based vector Lyapunov function approaches are discussed. The final chapter concludes with a survey of some recent results in this rather active area of research and some thoughts on problems still requiring a solution. In the Appendix we have derived the detailed and simplified models of the synchronous machine.

The book is aimed at a research audience as well as an application oriented power engineer who needs to understand the basis of direct methods of stability analysis in power systems. The book can be used in a graduate level course in power system stability or dynamics.

In writing this book, I have been influenced considerably by the work of many other research workers in this field whose works are cited. In particular, I would like to mention the report "Transient Energy Stability Analysis" by Systems Control Inc. in Proceedings of the Engineering Foundation Conference on 'System Engineering for Power: Emergency Operating State Control' sponsored by the U.S. Dept. of Energy at Davos, Switzerland in Oct. 1979. I also wish to thank Professor A. N. Michel of Iowa State University for numerous interesting discussions on the stability of large scale dynamical systems.

I would like to thank my former graduate students at I.I.T. Kanpur who got excited by this area of research and worked with great enthusiasm. My gratitude to I.I.T. Kanpur is an eternal one for nurturing an atmosphere of academic excellence and faculty support for research. I would like to thank the Council of Scientific and Industrial Research and the Dept. of Science and Technology, Govt. of India for supporting my research work at I.I.T. Kanpur.

The writing of this monograph was completed during my stay at Iowa State University during 1979-81. I am very grateful to Professor J. O. Kopplin, Chairman, Dept. of Electrical Engineering, for his constant encouragement and unfailing support. The support of the Engineering Research Institute of Iowa State University is also acknowledged. I would like to thank Shellie Siders for her speedy and excellent job of typing the manuscript.

Writing of this book has taken long hours away from my family, especially my son, Gurudutt, whom I thank for his understanding.

M.A. Pai

June 1981

CONTENTS

Chapter I

INTRODUCTION

1.1 Introduction

Any physical system that is designed or operated to perform certain preassigned tasks in a steady state mode must, in addition to performing these functions in a satisfactory manner, be stable at all times for sudden disturbances with an adequate margin of safety. When the physical system is large and complex such as a typical modern interconnected power system, investigation of stability requires both analytical sophistication in terms of techniques employed and practical experience in interpreting the results properly.

In the last decade and a half and in particular after the famous blackout in N.E. U.S.A. in 1965, considerable research effort has gone into the stability investigation of power systems both for off-line and on-line purposes. At the design stage although the planner takes many contingencies into consideration, in subsequent operation and augmentation of the network, new considerations arise which were not foreseen by the planner. Hence, an entirely different pattern of system behavior can be expected under actual operation conditions. This is particularly true regarding the capability of the system to maintain synchronism or stability due to sudden unforeseen disturbances such as loss of a major transmission line, load or generation. The tools suitable for off-line studies such as simulation may not be suitable for on-line application since a large number of contingencies have to be simulated in a short time. A technique which offers promise for this purpose is Lyapunov's method. The appeal of this method lies in its ability to compute directly the critical clearing time of circuit breakers for various faults and thus directly assess the degree of stability for a given configuration and operating state. The critical clearing time can also be translated in terms of additional power disturbances that the system can withstand, thus offering a tool for dynamic security assessment. For off-line applications, the method can serve as a complementary role to simulation by selecting quickly the important faults which need to be studied in detail. The state of art in this field resulting from nearly two decades of research work has reached a certain degree of maturity and for the first time offers possibilities of application by the electric utilities. The purpose of this book is to present Lyapunov's method and its systematic application to power system stability.

Perhaps for no other physical system has Lyapunov's stability theory been applied so extensively and vigorously as for power systems. It is difficult to precisely pinpoint historically the first application of Lyapunov's method for the investigation of power system stability. Since Lyapunov's function itself is a generalization of the energy function, one can say that Lyapunov's method indeed was applied implicitly to the power system problem when the equal area criterion was first proposed. The earliest work on applying the energy function to determine the stability of a power system was by Magnusson in 1947 [1]. An energy integral criterion as applied for multi-machine system was proposed by Aylett in 1958 [2]. Lyapunov's method as we know it in control literature today was first proposed as a solution to the power system stability problem by Gless [3] and El-Abiad and Nagappan [4]. In Ref. [4], the method was put on a firm footing as a powerful tool for stability analysis by power engineers through a realistic example of computing the critical clearing time and giving an algorithmic procedure for computer implementation. The early research such as those by Willems [5], Fallside and Patel [6], Dharma Rao [7], Yu and Vongasuria [8], and Siddiqee [9] concentrated on applying different methods of constructing Lyapunov functions for power systems. Di Caprio and Saccomanno [10] and Luders [11] related the physical aspects of the problem to the mathematical model and constructed an energy type Lyapunov function for multimachine systems. In 1970, the systematic construction of Lyapunov function to power systems using Popov's criterion was proposed by Willems [12], Willems and Willems [13], Pai, Mohan and Rao [14] and Pai [15]. It was soon realized that the more important aspect of applying the method to practical problems, namely the computation of critical clearing time, lay in computing accurately the region of attraction of the post-fault system. This has received considerable attention by Ribbens-Pavella beginning with her work in 1971 [16,17]. Since 1970, there has been an explosive growth in the research work on Lyapunov stability analysis of power systems. A number of state-of-the-art papers have appeared since then. In particular, the papers by Willems [18], Ribbens-Pavella [19], Saccomanno [20] and Fouad [21] summarize thoroughly the research efforts until 1975. Although the power industry evaluated the potential application of Lyapunov's method for large systems in 1972 [22], the method still had its drawbacks in terms of not giving consistently good results for critical clearing time for a wide range of fault conditions. Hence, largely theoretical aspects of the problem were pursued by the research workers in the 70's. Recently a significant breakthrough in obtaining excellent practical results has been achieved by Athay, Podmore and Virmani in U.S.A. [23], Ribbens-Pavella in Europe [24] and Kakimoto, Ohsawa and Hayashi in Japan [25]. Hence, what once appeared to be a mere academic exercise and viewed with skepticism by the power industry because of its conservative nature of results is today gaining acceptance both as an off-line tool for planning purposes as well as for on-line security assessment schemes. An alternative method of investigating stability of large scale power systems is through decomposition and subsequent aggregation. The concept of vector Lyapunov function proposed by Bailey [26] and extended to a wide class of composite systems by Michel and Porter [27], Michel and Miller [28], and Siljak [29,30] was first applied to power systems by Pai and Narayana [31].

Improvements in decomposition techniques in terms of reducing the number of subsystems as well as handling transfer conductances were proposed by Jocic, Ribbens-Pavella and Siljak [32]. However, results from a practical viewpoint are still conservative. In the future, this promises to be an area of active research and any breakthrough will have a major impact on the stability analysis of large scale power systems. There are two other promising research areas. One consists in stability analysis by retaining the topology of the transmission network [33] thus offering the possiblity of a better understanding of the phenomenon of loss of synchronism and also accomodating nonlinear load representation [34]. The other direction in which research is being pursued is incorporating accurate models of synchronous machines, excitation systems, etc. [35,36].

1.2 Statement of the Problem

The intuitive idea of stability of a physical system is as follows: Let the system be in some equilibrium state. If on the occurrence of a disturbance, the system eventually returns to the equilibrium position, we say the system is stable. The system is also termed stable if it converges to another equilibrium position generally in the proximity of the initial equilibrium point. If the state of the system "runs away" so that certain physical variables go on increasing as $t \to \infty$, then we say the system is unstable. These intuitive ideas are also applicable to a power system.

Suppose at a given time $t = t_o$, the power system is operating in a synchronous equilibrium and frequency equilibrium, i.e. the rotor angles of the different machines with respect to a rotating synchronous reference frame are fixed and the system frequency is constant. The constant nature of frequency implies that the system as a whole is at rest or the inertial center of the system is not accelerating. The occurrence of a disturbance tends to momentarily alter the synchronous equilibrium and the frequency equilibrium. The disturbance may either be small or big and the system may become unstable in either event depending on the operating condition at $t = t_o$. The study of stability in the presence of small disturbances constitutes what is known as "static stability" or "dynamic stability" analysis in the literature. The mathematical model for such a study is a set of linear time-invariant differential equations. When the disturbances are large, the nonlinearities inherent in the power system can no longer be ignored and the study of stability under such circumstances constitutes what is known as "transient stability" analysis. The mathematical model for such a study is a set of nonlinear differential equations coupled with a set of nonlinear algebraic equations. The dimensionality of the mathematical model both for dynamic and transient stability can be very large, for even moderately sized systems. We shall restrict ourselves to the transient stability aspects and now look at the physical phenomenon.

Assume that the system is in steady state with a frequency ω_o and the rotor angles δ_i^o of the various machines with respect to the synchronously rotating reference frame (i.e., $\delta_i = \alpha_i - \omega_o t$ where α_i is the angle with respect to a fixed reference) assuming constant values. Under these conditions, the power supplied by the generators exactly matches the power absorbed by the loads plus the electric power loss in the transmission system. At $t = t_o$, suppose a large perturbation in the system occurs, such as the loss of a major line or a fault at one of the buses. This disturbance upsets the energy balance existing prior to the disturbance. For example, a fault on a transmission line blocks the transfer of power to loads from certain nearby generators and will try to absorb power from other generators. This results in an excess or deficit of mechanical power supplied over the electrical power at each generator in the system resulting in acceleration or deceleration of the rotors. The rotor angles δ_i with respect to the synchronously rotating reference frame ω_o either increase or decrease as a function of time. A typical behavior or rotor angles is shown in Fig. 1.1. When the fault is removed by isolating the faulted line so that an energy balance is again possible, we have the problem of the redistribution of the excess kinetic energy acquired by the system until the instant of fault clearing. If the system after the fault is cleared can absorb this energy, the system is considered to be transiently stable; if the system cannot absorb all of this energy, then instability results. Figure 1.1 shows a system which is transiently stable and Fig. 1.2 an unstable system. When the system is transiently stable, it may recover to its original frequency or settle down to a new frequency. Knowledge regarding transient stability of the system is important both to the system planner and the operation engineer. The system planner has to coordinate the relay settings so that a given fault is cleared in time so as not to result in instability. A valuable piece of information in this respect is the critical clearing time t_{cr} of circuit breakers which isolates the faulty portion from the rest of the system. A fault cleared before t_{cr} will result in a stable system and a fault cleared after t_{cr} will result in the loss of synchronism. In order to compute t_{cr}, the system planner will have to conduct several stability runs each time fixing the clearing time t_{cl} and simulating the entire system. If the system is stable, t_{cl} is increased and the simulation is repeated. This is done until at some time when $t_{cl} = t_{cr}$, an inspection of the swing curves reveals instability. This is a tedious process besides being expensive in terms of computer time. Considering the fact that this procedure has to be repeated for different faults in the system, the need is evident for some direct method of computing t_{cr}. This is precisely sought to be done by Lyapunov's method.

In on-line applications, it is necessary to convert the information about the critical clearing time into some kind of a security index for interpretation by the operator. For example, the difference between the actual clearing time and the critical clearing time can be converted into a transient security index [37]. It is also possible to consider the effects of the outages on the transient stability. The application of Lyapunov's method to transient security monitoring is a relatively new area of research [38,39,40].

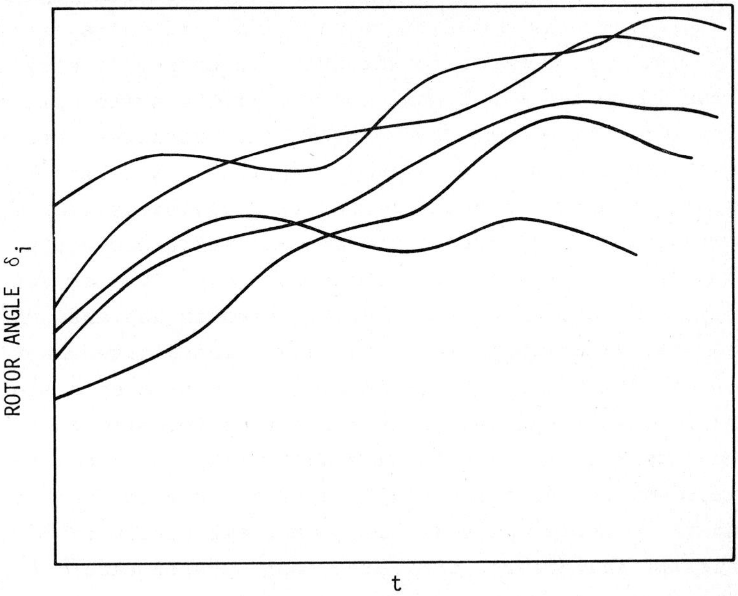

Fig. 1.1. Stable system.

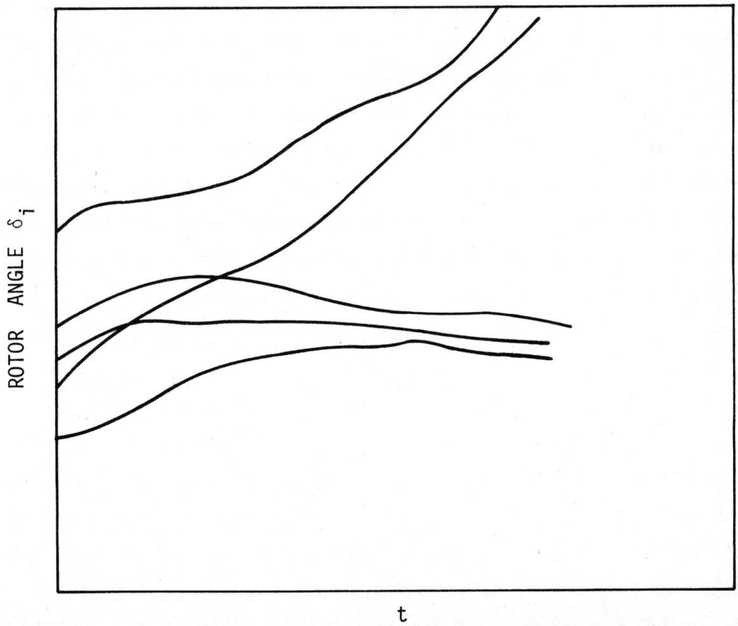

Fig. 1.2. Unstable system.

In Lyapunov's method, a certain scalar function of system state variables known as Lyapunov function $V(\underline{x})$ is computed first analytically. In many cases, the system energy is a good enough candidate for $V(\underline{x})$. We also compute a region of stability around the post fault stable equilibrium point denoted by the expression $V(\underline{x}) < V_{cr}$ where V_{cr} is the critical value of the Lyapunov function. In terms of energy, V_{cr} is the critical energy of the system which if exceeded during the fault-on period, the system will not be able to absorb it subsequently. There are many ways of computing V_{cr} and more recently, to get more accurate estimates of critical clearing time, the computation of V_{cr} is made fault dependent instead of being valid for all faults in the system for a given post-fault configuration. Also considerable research effort has been directed towards constructing better Lyapunov functions, accommodating accurate models of synchronous machines and extending the methodology for large scale systems through decomposition and vector Lyapunov function approach. All the collective research work done so far holds promise for Lyapunov's method being used as a tool for planning as well as on-line security assessment. As a planning tool, Lyapunov's stability analysis can be performed on a large number of faults to compute t_{cr} and consider for detailed study only those which are relatively less stable. In real time operation, the difference between V_{cr} and the value of $V(\underline{x})$ at $t = t_{cl}$ can be converted into stability margins for different contingencies and have the information displayed to the operator in a meaningful manner.

1.3 Chapter outline

We first give a basic review of Lyapunov stability theory in Chapter II. Various methods of constructing Lyapunov functions are covered. Lyapunov stability theory is discussed in slightly greater detail so as to serve as an independent and self-contained treatment of the subject. It is then followed in Chapter III by developing the mathematical model of multi-machine power systems. Both the center of angle reference frame and machine angle reference models are developed. The systematic construction of Lyapunov function for both single and multimachine models is then emphasized in Chapter IV using control theoretic as well as physical reasoning. For the single machine case, the region of attraction is computed. Computation of the region of stability for the multimachine case is the single biggest impediment to the method being used for on-line purposes. This is discussed in Chapter V. The transient energy function approach which takes into consideration the effect of faulted dynamics in the computation of V_{cr} and which has resulted in the accurate computation of t_{cr} is stressed. In Chapter VI, the vector Lyapunov function approach via decomposition is discussed. In the final chapter, some of the recent extensions and refinements in power system stability analysis by Lyapunov's method are briefly reviewed.

References

1. Magnusson, P. C., "Transient Energy Method of Calculating Stability", AIEE Trans., Vol. 66, 1947, pp. 747-755.

2. Aylett, P. D., "The Energy Integral-Criterion of Transient Stability Limits of Power Systems", Proceedings of IEE (London), Vol. 105C, No. 8, Sept. 1958, pp. 527-536.

3. Gless, G. E., "Direct Method of Lyapunov Applied to Transient Power System Stability", IEEE Trans. PAS, Vol. 85, No. 2, Feb. 1966, pp. 159-168.

4. El-Abiad, A. H. and Nagappan, K., "Transient Stability Regions of Multimachine Power Systems", IEEE Trans. PAS, Vol. 85, No. 2, Feb. 1966, pp. 169-178.

5. Willems, J. L., "Improved Lyapunov Function for Transient Power System Stability", Proceedings of IEE (London), Vol. 115, No. 9, Sept. 1968, pp. 1315-1317.

6. Fallside, F. and Patel, R., "On the Application of Lyapunov Method to Synchronous Machine Stability", International Journal of Control, Vol. 4, No. 6, Dec. 1966, pp. 501-513.

7. Rao, N. D., "Generation of Lyapunov Functions for the Transient Stability Problem", Trans. Engineering Institute of Canada, Vol. 11, Rep. C-3, Oct. 1968.

8. Yu, Y. N. and Vongasuriya, K., "Nonlinear Power System Stability Study by Lyapunov Function and Zubov's Method", IEEE Trans. PAS, Vol. 86, No. 12, Dec. 1967, pp. 1480-1485.

9. Siddiqee, F., "Transient Stability of AC Generator by Lyapunov's Direct Method", International Journal of Control, Vol. 8, No. 2, Aug. 1968, pp. 131-144.

10. Di Caprio, U. and Saccomanno, F., "Application of Lyapunov's Direct Method to the Analysis of Multimachine Power System Stability", PSCC Proceedings, Rome, Italy, 1969.

11. Luders, G. A., "Transient Stability of Multi-Power System via Direct Method of Lyapunov", IEEE Trans. PAS, Vol. 1, Jan./Feb. 1971, pp. 23-32.

12. Willems, J. L., "Optimum Lyapunov Functions and Stability Regions for Multi-Machine Power Systems", Proceedings of IEE (London), Vol. 117, No. 3, March 1970, pp. 573-578.

13. Willems, J. L. and Willems, J. C., "The Application of Lyapunov Methods to the Computation of Transient Stability Regions of Multi-Machine Power Systems", IEEE Trans. PAS, Vol. 89, No. 5, May/June 1970, pp. 795-801.

14. Pai, M. A., Mohan, M. A. and Rao, J. G., "Power System Transient Stability Region Using Popov's Method", IEEE Trans. PAS, Vol. 89, No. 5, May/June 1970, pp. 788-794.

15. Pai, M. A., "Power System Stability Studies by Lyapunov - Popov Approach", Proc. 5th IFAC World Congress, Paris, 1972.

16. Ribbens-Pavella, M. "Transient Stability of Multimachine Power Systems by Lyapunov's Method", IEEE Winter Power Meeting, New York, 1971.

17. Ribbens-Pavella, M. and Lemal, B., "Fast Determination of Stability Regions for On-line Transient Power System Studies", Proceedings of IEE (London), Vol. 123, No. 7, July 1976, pp. 689-696.

18. Willems, J. L., "Direct Methods for Transient Stability Studies in Power System Analysis", IEEE Trans. Automatic Control, Vol. A6-16, No. 4, Aug. 1971, pp. 332-341.

19. Ribbens-Pavella, M., "Critical Survey of Transient Stability Studies of Multi-Machine Power Systems by Lyapunov's Direct Method", Proceedings of 9th Annual Allerton Conference on Circuits and System Theory, Oct. 1971.

20. Saccomanno, F., "Global Methods of Stability Assessment", Proceedings of Symposium on Power System Dynamics, The University of Manchester Institute of Science and Technology, Manchester, England, Sept. 1973.

21. Fouad, A. A., "Stability Theory-Criteria for Transient Stability", Proc. Engineering Foundation Conference on "System Engineering for Power", Henniker, New Hampshire, 1975, pp. 421-450.

22. Willems, H. F., Louie, S. A. and Bills, G. W., "Feasibility of Lyapunov Functions for the Stability Analysis of Electric Power Systems Having up to 60 Generators", IEEE Trans. PAS, Vol. 91, No. 3, May/June 1972, pp. 1145-1153.

23. Athay, T., Podmore, R. and Virmani, S., "A Practical Method for Direct Analysis of Transient Stability", IEEE Trans. PAS, Vol. 98, No. 2, March/April 1979, pp. 573-584.

24. Ribbens-Pavella, M., Gruijc, Lj T., Sabatel, J. and Bouffioux, A., "Direct Methods for Stability Analysis of Large Scale Power Systems", Proc. IFAC Symposium on "Computer Applications in Large Scale Power Systems", New Delhi, India, Aug. 16-18, 1979, Pergamon Press, U.K.

25. Kakimoto, N., Ohsawa, Y., and Hayashi, M., "Transient Stability Analysis of Electric Power System vis Luré Type Lyapunov Function", Parts I and II, Trans. IEE of Japan, Vol. 98, No. 5/6, May/June, 1978.

26. Bailey, F. N., "The Application of Lyapunov's Method to Interconnected Systems", Journal SIAM (Control) Ser. A., Vol. 3, No. 3, 1966, pp. 443-462.

27. Michel, A. N. and Porter, D. W., "Stability of Composite Systems", Proc. Fourth Asilomar Conference on Circuits and Systems, Monterey, California, 1970.

28. Michel, A. N. and Miller, R. K., "Qualitative Analysis of Large Scale Dynamic Systems" (Book), Academic Press, New York, 1977.

29. Siljak, D. D., "Stability of Large Scale Systems", Proc. 5th IFAC World Congress, Paris, 1972.

30. Siljak, D. D., "Large Scale Dynamic Systems - Stability and Structure" (Book) North Holland, New York, 1978.

31. Pai, M. A. and Narayana, C. L., "Stability of Large Scale Power Systems", Proc. 6th IFAC World Congress, Boston, 1975.

32. Jocic, Lj B., Ribbens-Pavella, M., and Siljak, D. D., "On Transient Stability of Multi-Machine Power Systems", IEEE Trans. Automatic Control, Vol. AC-23, 1978, pp. 325-332.

33. Bergen, A. R. and Hill, D. J., "A Structure Preserving Model for Power Systems Stability Analysis", IEEE Trans. PAS, Vol. 100, Jan. 1981, pp. 25-35.

34. Athay, T. M. and Sun, D. I., "An Improved Energy Function for Transient Stability Analysis", Proc. International Symposium on Circuits and Systems, April 27-29, 1981, Chicago.

35. Sasaki, H., "An Approximate Incorporation of Field Flux Decay into Transient Stability Analysis of Multimachine Systems by the Second Method of Lyapunov", IEEE Trans. PAS, Vol. 98, No. 2, March/April 1979, pp. 473-483.

36. Kakimoto, N., Ohsawa, Y. and Hayashi, M., "Transient Stability Analysis of Multi-machine Power Systems with Field Flux Decays via Lyapunov's Direct Method", IEEE Trans. PAS, Vol. 99, No. 5, Sept./Oct. 1980, pp. 1819-1827.

37. Teichgraeber, R. D., Harris, F. W., and Johnson, G. L., "New Stability Measure for Multimachine Power Systems", IEEE Trans. PAS-89, Vol. 2, Feb. 1970, pp. 233-239.

38. Ribbens-Pavella, M., Calavaer, A. and Ghewry, A., "Transient Stability Index for On-line Evaluation", IEEE PES Winter Power Meeting, 1980.

39. Saito, O., Kinnosuke, K., Muneyuky, U., Masahiro, S., Hisao, M. and Toshihiko, T., "Security Monitoring System Including Fast Transient Stability Studies", IEEE Trans., PAS-94, Sept./Oct. 1975, pp. 1789-1805.

40. Fouad, A. A., Stanton, S. E., Mamandur, K.R.C., Kruempel, K. C., "Contingency Analysis Using the Transient Energy Margin Technique", Paper 81 SM 397-9, IEEE-PES Summer Power Meeting, Portland, Oregon, July 26-31, 1981.

Chapter II

REVIEW OF LYAPUNOV STABILITY THEORY

2.1 Introduction

The theory of stability of linear time-invariant systems is well known in the literature through the methods of Nyquist, Routh-Hurwitz, etc. which provide both necessary and sufficient conditions. However, in the case of nonlinear systems, no such systematic procedures exist. Closed-form solutions of nonlinear differential equations are exceptions rather than the rule. Although the digital computers can numerically integrate the differential equations, they do not provide an insight into the qualitative behavior of the differential equations. Consequently, the problem of finding the necessary and sufficient conditions for the stability of nonlinear systems is a formidable one and as yet an unsolved problem. All efforts in this direction relate directly or indirectly to the work of A. M. Lyapunov [1] who in 1892 set forth the general framework for the solution of such problems. Lyapunov dealt with both linear and nonlinear systems. He outlined two approaches to the problem of stability, known popularly as Lyapunov's "first method" and the other as the "second method of Lyapunov" or the "direct" method. The distinction is based on the fact that the "first method" depends on finding approximate solutions to the differential equations. In the "second method", no such knowledge is necessary. This is a great advantage in the case of nonlinear systems. We will be mainly concerned with the "second method of Lyapunov" or the "direct method" as applied to nonlinear systems.

A second major breakthrough in nonlinear stability theory came through the work of V. M. Popov [2] who in 1962 outlined a

$$\dot{V}(\underline{x}) = \dot{\underline{x}}^T \underline{P}\ \underline{x} + \underline{x}^T \underline{P}\ \dot{\underline{x}}$$

$$= \underline{x}^T (\underline{A}^T \underline{P} + \underline{P}\ \underline{A}) \underline{x}$$

$$= -\underline{x}^T \underline{Q}\ \underline{x}$$

Since $\underline{P}$ is positive definite, by Theorem 2.7.3 the origin is asymptotically stable if $\underline{Q} > 0$. Since origin is the only equilibrium point, the origin is also a.s.i.l. This leads us to a precise statement of the following theorem.

2.9.2 Stability Theorem for L.T.I. Systems

Suppose that $\underline{Q}$ is a positive definite symmetric matrix. Then for the system (2.7), $\underline{A}$ is a stable matrix, i.e. all eigenvalues of $\underline{A}$ have negative real parts if the solution for the real symmetric matrix $\underline{P}$ of the matrix equation

$$\underline{A}^T \underline{P} + \underline{P}\ \underline{A} = -\underline{Q} \tag{2.9}$$

is positive definite.

Furthermore, $V(\underline{x}) = \underline{x}^T \underline{P}\ \underline{x}$ constitutes a Lyapunov function for the system (2.7), having a negative definite time derivative

$$\dot{V}(x) = -\underline{x}^T \underline{Q}\ \underline{x}$$

Equation (2.9) is commonly known as Lyapunov's matrix equation.

We must clearly understand the implication of (2.9). Suppose we pick $\underline{Q}$ to be any p.d. matrix say $\underline{Q} = \underline{I}$, then if the equation has no solution or more than one solution, then the origin is not asymptotically stable. Suppose that (2.9) has a unique solution and $\underline{P}$ is not p.d., then again the origin is not asymptotically stable. On the other hand, if the unique solution is such that $\underline{P}$ is p.d., then the origin is asymptotically stable in the large.

The preceding theorem and discussion has not commented on what happens if $\underline{A}$ has eigenvalues on the imaginary axis. In terms of Lyapunov's stability theorem if there are distinct eigenvalues on the imaginary axis, the origin is stable but not criterion in the frequency domain for certain classes of non-linear control systems. This method is particularly attractive because of the ease with which it can be applied in much the same way as Nyquist's criterion is applied to linear systems. The mathematical model of power systems is amenable to analysis by Popov's method under certain simplifying assumptions.

A literature survey of the area of nonlinear stability using Lyapunov's method is practically impossible because of the vast amount of research surrounding this topic. There are a number of texts devoted to this topic constituting a good source of reference material on various aspects of stability theory (Hahn [3], LaSalle and Lefschetz [4], Willems [5], Vidyasagar [6]). What follows is a self-contained treatment of the topic with an emphasis on applications.

2.2 Mathematical Description of Systems

Most physical systems are described by their mathematical models for purposes of analysis. The following are some of the descriptions that are possible in continuous time systems. $\underline{x}$ represents the n dimensional state vector, $\underline{u}$ represents the r dimensional input vector and t represents the independent time variable.

1	$\dot{\underline{x}} = \underline{f}\ (\underline{x},\underline{u},t)$	Nonlinear, time varying and forced
2	$\dot{\underline{x}} = \underline{f}\ (\underline{x},t)$	Nonlinear, time varying and force free
3	$\dot{\underline{x}} = \underline{f}\ (\underline{x},\underline{u})$	Nonlinear, time invariant and forced
4	$\dot{\underline{x}} = \underline{f}\ (\underline{x})$	Nonlinear, time invariant and force free (autonomous)

The mathematical model of power systems for stability analysis results in an autonomous system of type (4) and fortunately for such type of systems a number of systematic procedures exist for constructing Lyapunov functions. In this chapter, we restrict the treatment to autonomous systems.

Consider the autonomous system

$$\dot{\underline{x}} = \underline{f}(\underline{x}) \tag{2.1}$$

i.e. $\dot{x}_1 = f_1(x_1, x_2, \ldots, x_n)$

$\dot{x}_2 = f_2(x_1, x_2, \ldots, x_n)$

$\cdot \quad \cdot$

$\cdot \quad \cdot$

$\dot{x}_n = f_n(x_1, x_2, \ldots, x_n)$

Let $\underline{x}_s$ be a solution of (2.1), i.e.

$$\dot{\underline{x}}_s = \underline{f}(\underline{x}_s) \tag{2.2}$$

If $\underline{x}_s$ = constant, then

$$\underline{0} = \underline{f}(\underline{x}_s) \tag{2.3}$$

Let at some instant of time, which can be designated as $t = 0$ without loss of generality, the system be disturbed instantaneously to a position $\underline{x}_0$ in the state space. The motion of the system for $t > 0$ is denoted by $\underline{x}(t)$. Define $\underline{y} = \underline{x} - \underline{x}_s$. Hence (2.1) becomes in view of (2.2)

$$\begin{aligned} \dot{\underline{y}} &= \underline{f}(\underline{x}_s + \underline{y}) - \underline{f}(\underline{x}_s) \\ &= \underline{g}(\underline{y}) \end{aligned} \tag{2.4}$$

with $\underline{y} = \underline{x}_0 - \underline{x}_s$

It is evident that $\underline{y} = \underline{0}$ satisfies (2.4) identically. Hence, instead of examining the stability of $\underline{x}_s$ in (2.1) we may equivalently examine stability of the origin of the equation $\dot{\underline{y}} = \underline{g}(\underline{y})$. This is often referred to as change of coordinates or transferring the equilibrium solution to the origin. Henceforth, we shall assume that this has been done and consider the stability of the origin (equilibrium point) of the equations

$$\dot{\underline{x}} = \underline{f}(\underline{x}), \quad \underline{f}(\underline{0}) = \underline{0} \tag{2.5}$$

The origin is equivalently called the trivial solution, singular point or the equilibrium point. For the sake of simplicity, let us assume that the origin is the only equilibrium point of (2.5). Furthermore, we will assume that (2.5) satisfies the so-called Lipschitz condition so that the solution of (2.5) exists and is unique for a given initial condition $\underline{x}_0$.

2.3 Definitions of Stability

We are all familiar with the intuitive idea of stability discussed in Chapter I. But precise definitions are needed before we can formulate the problem in more quantitative terms. In the mathematical literature on this subject, there are scores of definitions each differing from the other in a subtle way. However, for engineering purposes, we need to consider only about the stability of a certain motion of the system (2.1).

An undisturbed motion $\underline{x}_s$ of (2.1) satisfying (2.2) is considered to be stable if by giving a small disturbance to it, the disturbed motion remains close to the unperturbed one for all time. More specifically:

a) If for small disturbances, the effect on the motion is small, one says that the undisturbed motion is 'stable'.

b) If for small disturbances, the effect is considerable, the undisturbed motion is termed 'unstable'.

c) If for small disturbances, the effect tends to disappear, the undisturbed motion is 'asymptotically stable'.

d) If regardless of the magnitude of the disturbances, the effect tends to disappear, the undisturbed motion is 'asymptotically stable in the large'.

The above qualitative discussion will be put in more quantitative terms.

Consider the autonomous system

$$\dot{\underline{x}} = \underline{f}(\underline{x}) \tag{2.6}$$

where the origin representing the undisturbed motion is the equilibrium point in the state space, i.e.

$$\underline{0} = \underline{f}(\underline{0})$$

2.3.1 Stability

Definition 1: The origin is said to be stable in the sense of Lyapunov or simply stable if for every real number $\varepsilon > 0$ and initial time $t_0 > 0$ there exists a real number $\delta > 0$ depending on ε and in general on t such that for all initial conditions satisfying the inequality,

$$\|\underline{x}_0\| < \delta$$

the motion satisfies

$$\|\underline{x}(t)\| < \varepsilon \quad \text{for all } t > t_0.$$

The symbol $\|\cdot\|$ stands for the norm. In the Euclidean space, the norm can be interpreted to mean the Euclidean norm, i.e.,

$$\|\underline{x}\| = \left(\sum_{i=1}^{n} x_i^2\right)^{1/2}$$

Note that stability in the sense of Lyapunov is a local concept. It does not tell us apriori how small δ has to be chosen in the definition.

The origin is termed unstable if for a given ε as above and for sufficiently small δ there exists one solution which does not satisfy the inequality, $\|\underline{x}(t)\| < \varepsilon$.

The above definition is unsatisfactory from an engineering point of view since one is interested in a strong type of stability, namely that the solution must return to the origin as $t \to \infty$.

2.3.2 Asymptotic Stability

Definition 2: The origin is said to be asymptotically stable if it is stable and every motion starting sufficiently close to the origin converges to the origin as t tends to infinity, i.e.

$$\lim_{t \to \infty} \|\underline{x}(t)\| \to 0$$

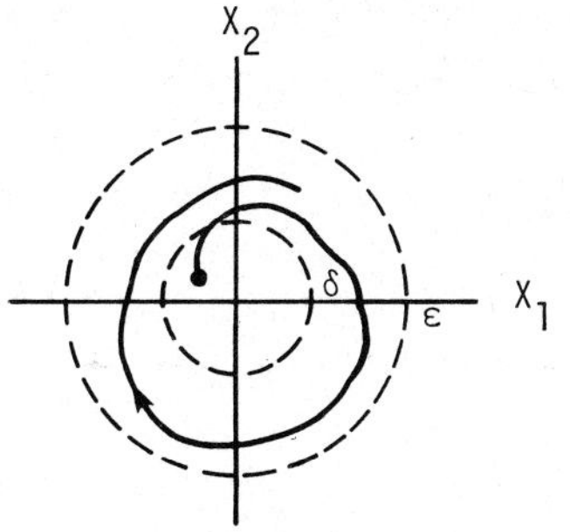

(a) Stability

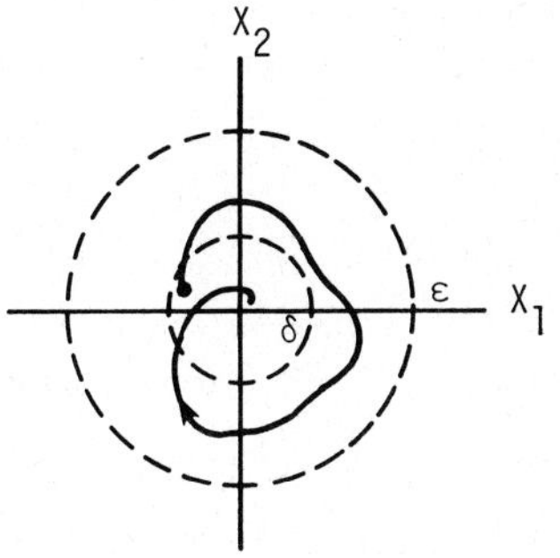

(b) Asymptotic stability

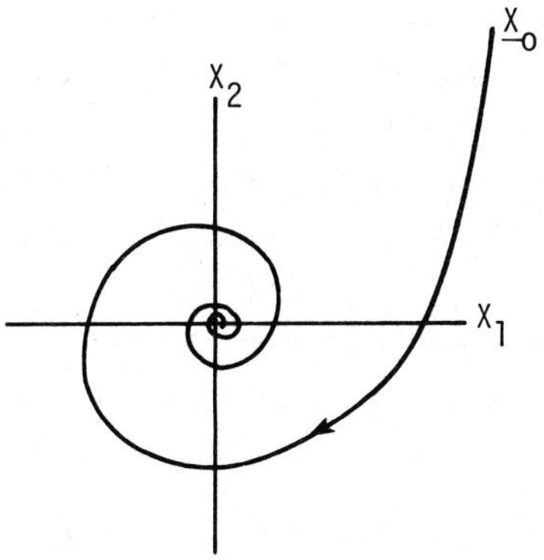

(c) Asymptotic stability in the large

Fig. 2.1. Various types of stability.

A deficiency of the above definition is that it does not indicate how big the disturbances can be in order that all subsequent motions will converge to the origin. In that sense, it is again a local concept.

2.3.3 Asymptotic Stability in the Large

Definition 3: The origin is said to be asymptotically stable in the large if it is asymptotically stable and every motion starting at any point in the state space, returns to the origin as t tends to infinity.

This is a very desirable type of stability in engineering systems since one does not have to worry about the magnitude of disturbances. The above three definitions can be illustrated geometrically as in Fig. 2.1 for a two-dimensional case.

2.3.4 Absolute Stability

Consider the nonlinear control system with zero input shown in Fig. 2.2(a). The nonlinearity is known to pass through the origin, lie within the first and third quadrants and is restricted to a region between the horizontal axis and a line of given slope k. Such a type of nonlinearity is said to satisfy a sector condition. The linear transfer function G(s) is fixed. If for the class of nonlinearities satisfying the stated restrictions, the origin is asymptotically stable in the large, we say that the nonlinear system is absolutely stable. For multiple nonlinearities, we can extend the concept to a transfer function matrix $\underline{W}(s)$ (Fig. 2.2(b)) with a vector nonlinearity $\underline{f}(\underline{\sigma})$ such $f_i(\underline{\sigma}) = f_i(\sigma_i)$ and each $f_i(\sigma_i)$ satisfies the sector condition.

2.4 Region of Asymptotic Stability

In many physical situations, the origin may not be asymptotically stable for all possible initial conditions but only for initial conditions contained in a region around the origin. Such a region is called the region of asymptotic stability or finite region of attraction. If such a region

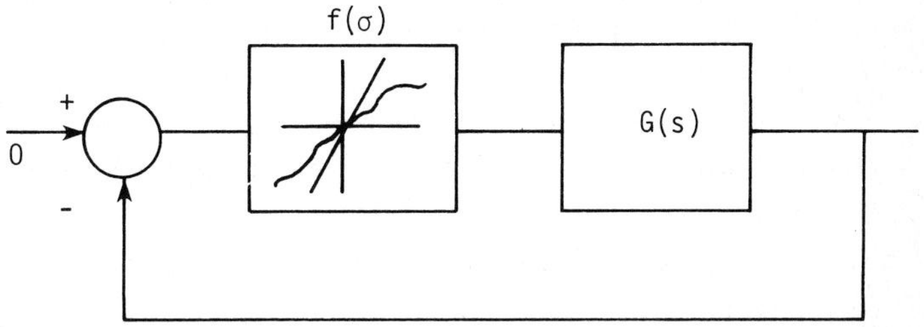

(a) Single nonlinearity

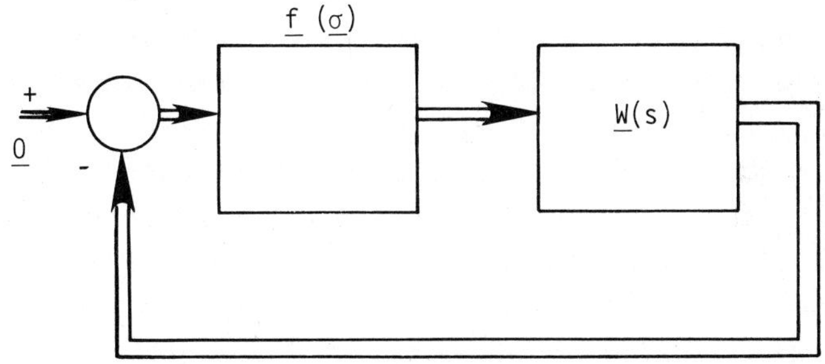

(b) Multiple nonlinearity with $f_i(\underline{\sigma}) = f_i(\sigma_i)$

Fig. 2.2. Single and multi-nonlinearity type feedback systems.

exists, computation of such a region is of great interest to the system designer. For example, in a multi-machine power system, if a fault is cleared too late, the post-fault system may not return to a stable state. This can be interpreted equivalently as the computation of a region of attraction around the post-fault stable state and determining if the state of the system at the point of clearing the fault lies within this region or not.

Examples:

i) Stability: Consider the harmonic oscillator equation.

$$\dot{x}_1 = x_2$$

$$\dot{x}_2 = -x_1$$

The origin is stable since given an ε within which we want the trajectory to lie, we can always give such an initial condition $\underline{x}_0$ lying within a circle of radius δ, $(\delta<\varepsilon)$ such that the trajectory lies within ε.

ii) Instability: Consider Vanderpol's equation $\ddot{x} + \varepsilon(x^2-1)x + x = 0$, $\varepsilon > 0$. The system has negative damping for small x and positive damping for large x. It has a stable limit cycle. The origin is unstable but the motion of the system is bounded.

iii) Stability: Consider the linear damped system.

$$\dot{x}_1 = x_2$$

$$\dot{x}_2 = -x_1 - x_2$$

The origin is asymptotically stable in the large.

iv) Region of Asymptotic Stability: Consider Vanderpol's equation with

$$t = -t$$

$$\ddot{x} - \varepsilon(x^2-1)\dot{x} + x = 0$$

The origin is asymptotically stable but not in the large. There is a finite region of asymptotic stability. The limit cycle is, however, unstable since a small perturbation around the limit cycle will take the trajectory away from the limit cycle.

2.5 Basis of Lyapunov's Method

The "direct" method of Lyapunov represents a philosophy of approach to the problem of stability and yet it is a very basic concept. The principle idea of the method is contained in the following reasoning. 'If the rate of change $\frac{dE}{dt}$ of the energy $E(\underline{x})$ of an isolated physical system is negative for every possible state $\underline{x}$ except for a single equilibrium state $\underline{x}_e$, then the energy will continually decrease until it finally assumes its minimum value $E(\underline{x}_e)$.' In our formulation $\underline{x}_e = \underline{0}$. The above concept was developed into a precise mathematical tool by Lyapunov. When the description of the system is given in a mathematical form, there is no way of defining energy. Hence, the energy function of $E(\underline{x})$ is replaced by some scalar function $V(\underline{x})$. If for a given system, one is able to find a function $V(\underline{x})$ such that it is always positive except at $\underline{x} = \underline{0}$ where it is zero and its derivative $\dot{V}(\underline{x})$ is < 0 except at $\underline{x} = \underline{0}$ where it is zero, then we say the system returns to the origin if it is disturbed. The function $V(\underline{x})$ is called the Lyapunov function. In what follows, we shall state somewhat more precisely Lyapunov's theorem on stability. Note that in the computation of $\dot{V}(\underline{x})$, the system equations come into the picture as shown below: Consider the autonomous equation

$$\dot{\underline{x}} = \underline{f}(\underline{x}),\ \underline{f}(\underline{0}) = \underline{0}$$

$$\dot{V}(x) = \frac{\partial V}{\partial x_1}\frac{dx_1}{dt} + \frac{\partial V}{\partial x_2}\frac{dx_2}{dt} \cdots \frac{\partial V}{\partial x_n}\frac{dx_n}{dt}$$

$$= \left\langle \frac{\partial V}{\partial \underline{x}}, \dot{\underline{x}} \right\rangle$$

$$= \langle \text{Grad } V, \dot{\underline{x}} \rangle = \langle \text{Grad } V, \underline{f}(\underline{x}) \rangle$$

2.6 Sign Definite Functions

The sign definite properties of a scalar function $V(\underline{x})$ will now be considered.

Definition 4: A function $V(\underline{x})$ is called positive definite (negative definite) if $V(\underline{0}) = 0$ and if around the origin $V(\underline{x}) > 0$ (<0) for $\underline{x} \neq 0$.

Definition 5: A function $\underline{V}(\underline{x})$ is called positive semi-definite (negative semi-definite) if $V(\underline{0}) = 0$ and if around the origin $V(\underline{x}) \geq 0$ (≤ 0) for $\underline{x} \neq 0$.

Examples:

i) For $n = 3$

$V(\underline{x}) = x_1^2 + x_2^2 + x_3^2$ is positive definite

ii) For $n = 3$

$V(\underline{x}) = x_1^2 + (x_2+x_3)^2$ is positive semi-definite since on $x_2 + x_3 = 0$ and $x_1 = 0$, $V(\underline{x}) = 0$

iii) For $n = 2$

$V(\underline{x}) = x_1^2 + x_2^2 - (x_1^4+x_2^4)$ is positive definite near the origin. In general, it is sign indefinite. More precisely, it is positive definite inside the square defined by $|x_1| < 1$, $|x_2| < 1$.

2.7 Stability Theorems

In this section, we state the theorems on stability of the origin of the autonomous equation defined by (2.6).

2.7.1 Theorem on Stability

The origin of (2.6) is said to be stable if there exists a scalar function $V(\underline{x}) > 0$ in the neighborhood of the origin such that $\dot{V}(\underline{x}) \leq 0$ in that region. Such a function is known as the Lyapunov function.

Example:

Consider the equation corresponding to the harmonic oscillator.

$$\dot{x}_1 = x_2$$

$$\dot{x}_2 = -x_1$$

The origin is the only equilibrium point as can be seen by inspection. Choose $V(\underline{x}) = x_1^2 + x_2^2$ which is positive definite.

$$\begin{aligned}\dot{V}(\underline{x}) &= \frac{\partial V}{\partial x_1}\frac{dx_1}{dt} + \frac{\partial V}{\partial x_2}\frac{dx_2}{dt} \\ &= 2x_1x_2 - 2x_1x_2 \\ &= 0\end{aligned}$$

Hence, the condition

$$\dot{V}(\underline{x}) \leq 0 \text{ is satisfied and } V(\underline{x}) > 0$$

Therefore, the origin is stable. If we associate the physical system with the differential equations, namely, the harmonic oscillator, it is easily verified that $V(\underline{x})$ in fact represents the total energy of the system. The fact that $\dot{V}(\underline{x}) = 0$ merely indicates that it is a conservative system and the total energy remains constant.

Example:

Consider the system of equations

$$\dot{x}_1 = x_2 - 3x_3 - x_1(x_2-2x_3)^2$$

$$\dot{x}_2 = -2x_1 + 3x_3 - x_2(x_1+x_3)^2$$

$$\dot{x}_3 = 2x_1 - x_2 - x_3$$

Choose

$$V(\underline{x}) = 2x_1^2 + x_2^2 + 3x_3^2$$

$$\dot{V}(\underline{x}) = -4x_1^2(x_2-2x_3)^2 - 2x_2^2(x_1+x_3)^2 - 6x_3^2$$

$\dot{V}(\underline{x})$ is negative semi-definite, being zero for $x_2 = x_3 = 0$ and $x_1 \neq 0$. The origin is therefore stable.

2.7.2 Theorem on Asymptotic Stability

The origin of (2.6) is asymptotically stable if there exists a scalar function $V(\underline{x}) > 0$ such that $\dot{V}(\underline{x}) < 0$. If such a function exists, it is called a Lyapunov function for the system or simply the V function.

Example:

Consider the system of equations

$$\dot{x}_1 = x_2 - ax_1(x_1^2+x_2^2)$$

$$\dot{x}_2 = -x_1 - ax_2(x_1^2+x_2^2)$$

where 'a' is a positive constant. Choose

$$V(\underline{x}) = x_1^2 + x_2^2$$

Then $$\dot{V}(x) = -2a(x_1^2+x_2^2)^2 = -2a\ V^2(\underline{x})$$

Hence, $V(\underline{x})$ constitutes a Lyapunov function for the system and the origin is asymptotically stable.

Example:

$$\dot{x}_1 = (x_1-\beta x_2)(ax_1^2+bx_2^2-1)$$

$$\dot{x}_2 = (\alpha x_1+x_2)(ax_1^2+bx_2^2-1)$$

Choose

$$V(\underline{x}) = \alpha x_1^2 + \beta x_2^2$$

$$\dot{V}(\underline{x}) = 2(\alpha x_1^2+\beta x_2^2)(ax_1^2+bx_2^2-1).$$

If a, b, α, β are all positive, then $V(\underline{x})$ is positive definite and $\dot{V}(\underline{x})$ is negative definite in the domain $ax^2 + bx^2 < 1$. The origin is asymptotically stable for all initial conditions in this domain. Note that in this instance by examining $\dot{V}(\underline{x})$ we are able to get an estimate of the domain of attraction. The exact region of asymptotic stability is bigger than this since Lyapunov's theorem gives only sufficient but not necessary and sufficient conditions.

2.7.3 Theorem on Asymptotic Stability in the Large

The origin of (2.6) is asymptotically stable in the large if there exists a scalar function $V(\underline{x})$ such that

i) $V(\underline{x}) > 0$

ii) $\dot{V}(\underline{x}) < 0$

iii) $V(\underline{x}) \to \infty$ as $||\underline{x}|| \to \infty$ i.e., $V(\underline{x})$ is radially unbounded.

Condition (iii) ensures that $V(\underline{x}) = C$ (C is a constant) represents closed surfaces for all C in the entire state space. An example of a radially unbounded function is

$$V(\underline{x}) = x_1^2 + x_2^2 \text{ whereas } V(\underline{x}) = \frac{x_1^2 + 2x_2^2}{x_1^2 + 2}$$

is not radially unbounded. Theorem 2.7.3 is difficult to apply in practical situations because it is difficult to find $V(\underline{x}) > 0$ with $\dot{V}(\underline{x}) < 0$ in the entire state space. Hence, we have an alternative theorem.

2.7.4 Alternative Theorem on Asymptotic Stability in the Large

In Theorem 2.7.3, Condition (ii) can be replaced by

ii) (a) $\dot{V}(\underline{x}) \leq 0$

ii) (b) $\dot{V}(x)$ does not vanish identically for $t \geq 0$ along any other solution excepting $\underline{x} = 0$. Stated in another manner, (ii) (b) implies that if $\dot{V}(\underline{x}) = 0$ at points other than the origin, then these points should not constitute a solution of the system.

Example:

Consider the series RLC circuit. The differential equation is

$$iR + L\frac{di}{dt} + \frac{1}{C}\int i\,dt = 0$$

If R is nonlinear resistor, then $R = f(i)$

$$i\,f(i) + L\frac{di}{dt} + \frac{1}{C}\int i\,dt = 0$$

Choosing $q = x_1$ and $\dot{q} = i = x_2$ as the state variables, the system equations can be written as

$$\dot{x}_1 = x_2$$

$$\dot{x}_2 = -\frac{1}{LC}x_1 - \frac{1}{L}x_2\,f(x_2)$$

Suppose we choose energy as the Lyapunov function

$$V(x_1,x_2) = \frac{1}{2C}x_1^2 + \frac{1}{2}L\,x_2^2$$

Then

$$\dot{V}(x_1,x_2) = \frac{x_1}{C}\dot{x}_1 + L\,x_2\dot{x}_2$$

$$= \frac{x_1x_2}{C} + L\,x_2\left(-\frac{1}{LC}x_1 - \frac{1}{L}x_2\,f(x_2)\right)$$

$$= -x_2^2\,f(x_2)$$

Assume positive values of resistance only so that $f(x_2) > 0$. Hence, $V(\underline{x}) > 0$ and $\dot{V}(\underline{x}) \leq 0$. Also $V(\underline{x}) \to \infty$ as $||\underline{x}|| \to \infty$. We now invoke the alternative formulation of the theorem on a.s.i.l. and verify that $V(\underline{x})$ does not vanish along any solution other than the origin. $\dot{V}(\underline{x}) = 0$ for $x_2 = 0$ but x_1 variable i.e. the x_1 axis. But it does not constitute a solution of the system or a trajectory except $x_1 = x_2 = 0$. Therefore, we conclude that the origin is a.s.i.l.

In many physical systems, it is much easier to find $V(\underline{x}) > 0$ and $\dot{V}(\underline{x}) \leq 0$ rather than a strict negative definite $\dot{V}(\underline{x})$. Hence, Theorem 2.7.4 will be found to be most readily amenable for engineering applications. The fact that $\dot{V}(\underline{x}) \neq 0$ along any solution other than origin should be argued out from physical considerations as in the preceding example.

2.7.5 Theorem on Region of Asymptotic Stability

Consider the autonomous equation

$$\dot{\underline{x}} = \underline{f}(\underline{x}), \; \underline{f}(\underline{0}) = \underline{0}$$

Let $V(\underline{x})$ be a scalar function. Let Ω designate a region where $V(\underline{x}) > 0$. Assume that Ω is bounded and that within

i) $V(\underline{x}) > 0$

ii) $\dot{V}(\underline{x}) < 0$

Then the origin is asymptotically stable and all motions starting in Ω converge to the origin as $t \to \infty$.

Condition (ii) can be replaced by (ii)a and (ii)b as in Theorem 2.7.4.

From the above theorem, a rule can be constructed for finding an estimate of the region of asymptotic stability as follows: Choose a $V(\underline{x})$ for all $\underline{x}$ and let $\dot{V}(\underline{x})$ be < 0 near the origin. Let V_{min} be the lowest value of $V(\underline{x})$ on the surface $\dot{V}(\underline{x}) = 0$. The region determined by

$$V(\underline{x}) < V_{min}$$

is contained in the region of asymptotic stability and therefore provides an estimate of this region. Note that the region obtained by the inequality given above depends on the choice of $V(\underline{x})$ and different choices of $V(\underline{x})$ may yield different estimates of the region of asymptotic stability. Then the union of these regions is contained in the exact region of asymptotic stability. Thus, improved estimates can be obtained by taking different Lyapunov functions.

Example:

Consider the equations

$$\dot{x}_1 = -3x_2 - x_1 + x_1^3$$

$$\dot{x}_2 = -2x_2 + x_1 - x_1^3$$

Choose $V(\underline{x}) = 2x_1^2 - 2x_1x_2 + 3x_2^2$

It is easily verified by applying Sylvester's theorem that $V(\underline{x}) > 0$.

$$\dot{V}(\underline{x}) = -6(x_1^2 - x_1^4 + \frac{4}{3} x_1^3 x_2 + x_2^2)$$

Near the origin $\dot{V}(\underline{x}) < 0$ but in general $\dot{V}(\underline{x}) \leq 0$. We proceed to find the region of asymptotic stability by examining the surface $\dot{V}(\underline{x}) = 0$. It is sketched in Fig. 2.3. Setting $x_2 = 0$, it is seen that $\dot{V}(\underline{x}) = 0$ for $x_1 = \pm 1$. Computations reveal the curves $\dot{V}(\underline{x}) = 0$ as open surfaces in $x_1 - x_2$ plane. We must find V_{min} on this surface. Computationally this can be quite difficult. However, we see that the surface defined by $V(\underline{x}) < 2$ lies in the region $\dot{V} < 0$ and is shown in the figure. It is therefore an estimate of the region of asymptotic stability.

Example:

Consider Vanderpol's equation

$$\ddot{x} + \varepsilon(x^2-1)\dot{x} + x = 0, \quad \varepsilon > 0$$

Define the state variables as

$$x_1 = \dot{x} + \int_o^x \varepsilon(u^2-1)du$$

$$x_2 = x$$

Then $\dot{x}_1 = -x_2$

$$\dot{x}_2 = x_1 - \varepsilon(\frac{x_2^3}{3} - x_2)$$

The above equations are known to have a stable limit cycle. The equilibrium point (origin) is an unstable point. If we put $t = -t$, then the origin will become a stable equilibrium point with an exact region of asymptotic stability given by the limit cycle. By using Theorem 2.7.5, we shall get an estimate of this region. Putting $t = -t$

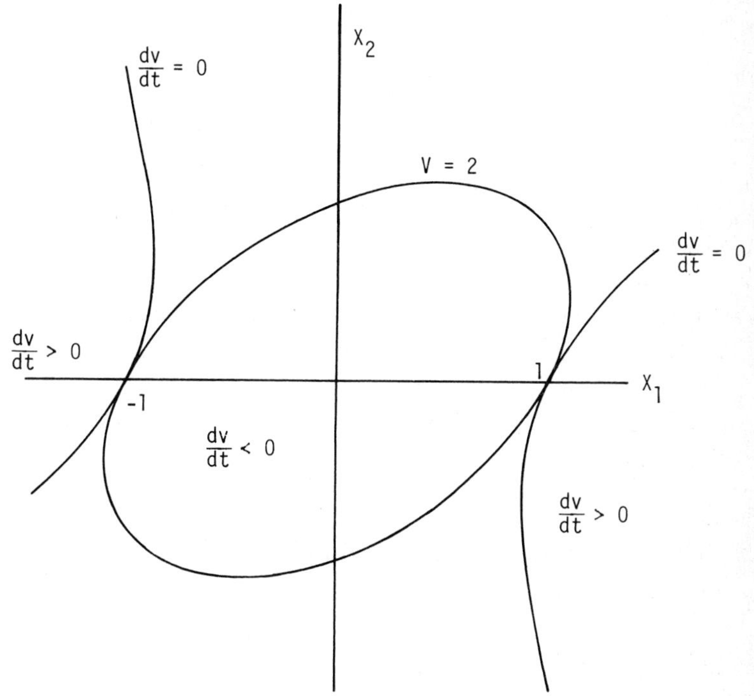

Fig. 2.3. Computing region of asymptotic stability.

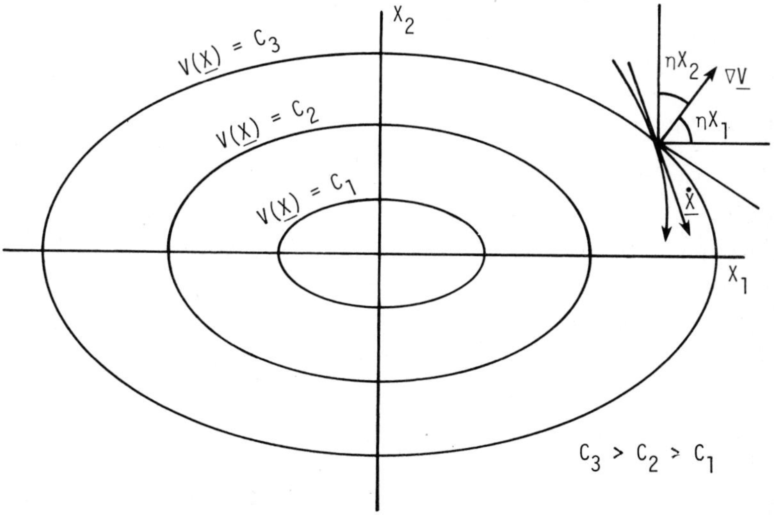

Fig. 2.4. Geometrical interpretation of the stability criterion.

$$\dot{x}_1 = x_2$$

$$\dot{x}_2 = -x_1 + \varepsilon(\frac{x_2^3}{3} - x_2)$$

Choose

$$V(\underline{x}) = \frac{x_1^2 + x_2^2}{2}$$

$$\dot{V}(\underline{x}) = x_1\dot{x}_1 + x_2\dot{x}_2$$

$$= x_1x_2 + x_2(-x_1 + \varepsilon(\frac{x_2^3}{3} - x_2))$$

$$= \varepsilon x^2(\frac{x_2^2}{3} - 1)$$

$$\dot{V}(\underline{x}) \leq 0 \text{ for } x_2^2 < 3$$

$\dot{V}(\underline{x}) = 0$ on the x_1 axis. On the entire x_1 axis $\frac{dx_2}{dx_1} = \pm \infty$ and hence the x_1 axis for $x_1 \neq 0$ does not constitute a solution of the system. Hence the only solution is the origin. Since $\dot{V}(\underline{x}) \leq 0$ for $|x_2| < \sqrt{3}$ the value of V_{min} is evaluated by computing $V(\underline{x})$ at $x_1 = 0$, $x_2 = \pm \sqrt{3}$. This is easily found to be 3/2. Hence an estimate of the region of asymptotic stability is given by

$$V(\underline{x}) = \frac{x_1^2 + x_2^2}{2} < 3/2$$

or $$x_1^2 + x_2^2 < 3$$

This corresponds to a circle of radius $\sqrt{3}$ and the actual limit cycle, whatever the value of ε, lies exterior to this. The conservative nature of the region of asymptotic stability is both due to the particular choice of the Lyapunov function and its sufficiency nature.

Example:

Consider the equations

$$\dot{x}_1 = x_2$$

$$\dot{x}_2 = -2b\, x_1 - ax_2 - 3x_1^2$$

We will examine the stability of the origin and try to construct a region of asymptotic stability. Take $V(\underline{x})$ as

$$V(\underline{x}) = bx_1^2 + x_1^3 + \frac{x_2^2}{2}$$

$V(\underline{x}) > 0$ in a region around the origin. In fact, it is indefinite in sign for values of $x_1 < -b$. $\dot{V}(\underline{x})$ is evaluated as

$$\dot{V}(\underline{x}) = -ax_2^2$$

$\dot{V}(\underline{x}) \leq 0$ around the origin. The set of points E for which $\dot{V}(\underline{x}) = 0$ is the entire x_1 axis. Examining

$$\frac{dx_2}{dx_1} = \frac{-2bx_1 - ax_2 - 3x_1^2}{x_2}$$

it is ∞ outside the origin and zero at the point (-2b/3,0) and (0,0). Hence, the equilibrium points are the origin and the point (-2b/3,0). Evaluating $V(\underline{x})$ at the point (-2b/3,0), we get the region $V(\underline{x}) < \frac{4b^3}{27}$ defined as an estimate of the region of asymptotic stability of the origin.

From the preceding three examples, we see that the region of attraction $V(\underline{x}) < C$ can be computed either by evaluating C on the curve $\dot{V}(\underline{x}) = 0$ and taking minimum value or evaluating $V(\underline{x})$ at the nearest equilibrium point other than the origin. As it turns out, the latter method has been used extensively in power systems.

2.8 Proofs of Stability Theorems

We have so far only stated the theorems with several illustrative examples. The proofs have been omitted since they involve a reasonable amount of mathematical rigor. However, a geometric interpretation of some of the theorems is found to be of great value. Consider

$$\frac{dV}{dt} = \sum_{k=1}^{n} \frac{\partial V}{\partial x_k} \cdot \dot{x}_k$$

Consider the surface $V(\underline{x}) = C$, $C > 0$ and some point on the surface. The quantities $\frac{\partial V}{\partial x_k}$ are the direction derivatives along the different axis. The normal to the curve $V(\underline{x})$ is given by the vector $\underline{\nabla V}$ whose components are $\frac{\partial V}{\partial x_k}$. If ηx_k represents angle of the normal to x_k axis, then

$$\frac{\partial V}{\partial x_k} = |\underline{\nabla V}| \cos \eta x_k$$

The positive direction of the normal is in the direction of increasing $V(\underline{x})$. Now

$$\frac{dV}{dt} = \sum_{k=1}^{n} \frac{\partial V}{\partial x_k} \frac{dx_k}{dt} = |\underline{\nabla V}| \sum_{k=1}^{n} \frac{dx_k}{dt} \cos \eta x_k$$

The term $\sum_{k=1}^{n} \frac{dx_k}{dt} \cos \eta x_k$ represents the sum of the projections of the velocity vector components on the normal $\underline{\nabla V}$. If $\frac{dV}{dt} < 0$, it means trajectories move towards surfaces with smaller value of C. This in turn implies that the trajectories intersect the families of surfaces inward since surfaces for small values of C are contained in surfaces with greater values of C. If $dV/dt > 0$, then the trajectories cross $V(\underline{x}) = C$ surfaces from inside. If $\sum_{k=1}^{n} \frac{dx_k}{dt} \cos \eta x_k = 0$, the trajectory lies on the surface $V(\underline{x}) = C$. The preceding observations can be verified from Fig. 2.4 for a two-dimensional case.

2.9 Lyapunov Functions for Linear Time Invariant Systems

In this section, we will consider the stability investigation of linear time invariant (L.T.I.) systems by Lyapunov's method.

Consider the system

$$\dot{\underline{x}} = \underline{A}\,\underline{x} \tag{2.7}$$

with origin as the only equilibrium point whose stability is to be investigated. The stability of (2.7) can be examined either by computing the eigenvalues of $\underline{A}$ to see if any of them lie in the right half plane or not. Alternatively, the Routh Hurwitz conditions can be applied to the characteristic equation given by $\det[\underline{A} - \lambda\underline{I}] = 0$. However, both these methods do not readily give an insight into the class of $\underline{A}$ matrices which are stable. This is possible through the construction of a Lyapunov function of a quadratic form which also yields the necessary and sufficient conditions regarding stability of the $\underline{A}$ matrix.

The following theorem regarding the solution of a matrix equation is of fundamental importance.

Consider the matrix equation

$$\underline{A}^T\underline{P} + \underline{P}\,\underline{A} = -\underline{Q} \tag{2.8}$$

where $\underline{A}$ is the given n×n matrix, $\underline{P}$ and $\underline{Q}$ are symmetric n×n matrices. If $\underline{Q}$ is given, the above matrix equation results in a set of $\frac{n(n+1)}{2}$ linear equations for the elements of $\underline{P}$.

Theorem: If $\lambda_1, \lambda_2, \ldots, \lambda_n$ are the eigenvalues of the matrix $\underline{A}$, then the above equation has a unique solution for $\underline{P}$ if and only if

$$\lambda_i + \lambda_j \neq 0 \quad \text{for all} \quad i, j = 1,2,\ldots,n$$

This implies that for a unique solution to exist, $\underline{A}$ should have no zero eigenvalues and no real eigenvalues which are of opposite sign. The proof of this theorem is contained in Hahn [3].

2.9.1 Lyapunov function for L.T.I. Systems

We choose $V(\underline{x}) = \underline{x}^T\underline{P}\,\underline{x}$ where $\underline{P}$ is positive definite, as the Lyapunov function for (2.7). Then asymptotically stable. The question is partly answered by the following theorem due to B. D. O. Anderson [7].

Theorem: If the matrix $\underline{A}$ has no eigenvalues with positive real parts and has some eigenvalues with zero real parts which are distinct, then for a given $\underline{Q} \geq 0$, there exists a $\underline{P} > 0$ satisfying the Lyapunov matrix equation (2.9). Furthermore, the system is stable with $V(\underline{x}) = \underline{x}^T \underline{P}\, \underline{x}$ and $\dot{V}(\underline{x}) = -\underline{x}^T \underline{Q}\, \underline{x} \leq 0$.

2.10 Construction of Lyapunov Functions for Nonlinear Systems

One of the main impediments to the application of Lyapunov's method to physical systems is the lack of formal procedures to construct the Lyapunov function for the differential equations describing the given physical system. We have seen in the previous section that for a linear time invariant system, systematic procedures exist via the Lyapunov matrix equation. However, the bulk of practical problems fall under the nonlinear category for which systematic procedures are an exception rather than the rule. A number of investigators have worked on evolving some kind of general methods applicable to a certain class of problems. A good categorization and historical summary of various methods of constructing Lyapunov functions is contained in a paper by Gurel and Lapidus [8]. We can broadly categorize the methods as

1) Methods based on first integrals
2) Methods based on quadratic forms
3) Methods based on solution to partial differential equation
4) Methods based on quadratic plus integral of nonlinearity type Lyapunov function
5) Miscellaneous methods

2.11 Methods Based on First Integrals [9]

This method of constructing Lyapunov functions is based on a linear combination of first integrals of the system equations. In turn, this is based on Lyapunov's original idea that total energy, in the case of a conservative system, could serve to define the stability of an equilibrium point.

We shall first define what we mean by a first integral. Consider

$$\dot{x}_i = f_i(x_1, x_2, \ldots, x_n)$$
$$\text{i.e.} \quad \underline{\dot{x}} = \underline{f}(\underline{x}), \; \underline{f}(\underline{0}) = \underline{0} \tag{2.10}$$

By a first integral (or simply integral) we understand a differentiable function $G(x_1, x_2, \ldots, x_n)$ defined in domain D of the state space such that when x_i's constitute a solution, $G(x_1, x_2, \ldots, x_n)$ assumes a constant value C. Existence of a first integral can be used to define a conservative system. A necessary and sufficient condition for (2.10) to have a first integral is given by the condition

$$\sum_{i=1}^{n} \frac{\partial f_i}{\partial x_i} = 0 \tag{2.11}$$

There are no general methods to construct the first integral except for second order systems.

Example:

Consider

$$\dot{x}_1 = x_2$$
$$\dot{x}_2 = 4x_1^3 - 4x_1 \tag{2.12}$$

It is easily verified that condition (2.11) is satisfied for this equation. Now (2.12) can be manipulated as

$$\frac{dx_1}{dx_2} = \frac{x_2}{4x_1^3 - 4x_1}$$

Multiplying and integrating we get

$$G(x_1, x_2) = \frac{x_2^2}{2} + \int_0^{x_1} 4(x_1 - x_1^3)\,dx_1$$

as the first integral or energy integral of the system. Since $\dot{G}(x_1, x_2) = 0$, and $G(x_1, x_2)$ is positive definite around the origin, it constitutes a Lyapunov function. The origin is stable.

The method of first integrals is limited to systems which are conservative and which in addition satisfy condition (2.11).

2.12 Method of Quadratic forms (Krasovskii's Method) [10]

In this type of method, the Lyapunov function is of the form $\underline{x}^T\underline{A}(\underline{x})\underline{x}$ or $\underline{f}^T\underline{P}(\underline{x})\underline{f}$. Consider the autonomous system

$$\dot{\underline{x}} = \underline{f}(\underline{x}), \quad \underline{f}(\underline{0}) = 0 \tag{2.13}$$

Assume that $\underline{f}(\underline{x})$ has continuous first partial derivatives and define the Jacobian matrix

$$\underline{J}(\underline{x}) = \frac{\partial \underline{f}}{\partial \underline{x}} = \begin{bmatrix} \frac{\partial f_1}{\partial x_1} & \cdots & \frac{\partial f_1}{\partial x_n} \\ \vdots & & \vdots \\ \frac{\partial f_n}{\partial x_1} & \cdots & \frac{\partial f_n}{\partial x_n} \end{bmatrix} \tag{2.14}$$

Krasovskii's Criterion

If there exists a constant positive definite matrix $\underline{P}$ such that the matrix $\underline{Q}(\underline{x})$ defined by

$$\underline{Q}(\underline{x}) = \underline{P}\,\underline{J}(\underline{x}) + \underline{J}^T(\underline{x})\underline{P}$$

is negative definite, then the origin of (2.13) is asymptotically stable in the large.

Consider the Lyapunov function

$$V(\underline{x}) = \underline{f}^T\underline{P}\,\underline{f}$$

which is positive definite in the $\underline{f}$ space. Since there is a one to one mapping between $\underline{x}$-space and $\underline{f}$-space, $V(\underline{x})$ is also positive definite in the $\underline{x}$-space. The derivative of $V(\underline{x})$ is given by

$$\dot{V}(\underline{x}) = \dot{\underline{f}}^T\underline{P}\,\underline{f} + \underline{f}^T\underline{P}\,\dot{\underline{f}}$$

By chain rule

$$\begin{aligned}\dot{f}(x) &= J(x)\dot{x} \\ &= \underline{J}(\underline{x})\underline{f}(\underline{x})\end{aligned}$$

Hence

$$\dot{\underline{V}}(\underline{x}) = \underline{f}^T[\underline{J}^T(\underline{x})\underline{P} + \underline{P}\ \underline{J}(\underline{x})]\underline{f}$$

$\dot{V}(\underline{x})$ is negative definite since the term inside the brackets is negative definite. Hence, the origin is asymptotically stable in the large.

Example:

$$\dot{x}_1 = -ax_1 + x_2 \qquad\qquad a > 1 \qquad\qquad (2.15)$$
$$\dot{x}_2 = x_1 - x_2 - x_2^3$$

which has $\underline{0}$ as the only equilibrium point

$$\underline{J}(\underline{x}) = \begin{bmatrix} -a & 1 \\ 1 & -1 - 3x_2^2 \end{bmatrix}$$

Choosing $\underline{P} = \underline{I}$, we have $\underline{J}^T(\underline{x}) + \underline{J}(\underline{x}) = \underline{Q}(\underline{x})$ where

$$\underline{Q}(\underline{x}) = \begin{bmatrix} -2a & 2 \\ 2 & -2 - 6x_2^2 \end{bmatrix}$$

The condition for a.s.i.l. of the origin is that $\underline{Q}(\underline{x})$ be negative definite i.e., $-2a < 0$ and $4a + 12ax_2^2 - 4 > 0$. Since $a > 1$, both these inequalities are satisfied. Hence, the origin is a.s.i.l.

2.13 Variable Gradient Method [11]

As the name implied, this method is based on the assumption of a vector $\nabla\underline{V}$ with n undetermined components. An important feature of the gradient function is that both $\dot{V}$ and V may be determined from it. Thus

$$\frac{dV}{dt} = \frac{\partial V}{\partial x_1}\dot{x}_1 + \ldots . \frac{\partial V}{\partial x_n}\dot{x}_n$$

$$= \langle\underline{\nabla V}, \dot{\underline{x}}\rangle$$

and $V = \int_0^{\underline{x}} \langle \nabla \underline{V}, d\underline{x} \rangle$

The upper limit $\underline{x}$ in the integral indicates that the line integral is to an arbitrary point in the $\underline{x}$ space and is independent of the path of integration. The explicit form is

$$V = \int_0^{x_1} \nabla V_1(\gamma_1, 0 \ldots 0)\, d\gamma_1 + \int_0^{x_2} \nabla V_2(x_1, \gamma_2, 0 \ldots 0)\, d\gamma_2 + \ldots \int_0^{x_n} \nabla V_n(x_1, x_2, \ldots, x_{n-1}, \gamma_n)\, d\gamma_n \tag{2.16}$$

where ∇V_i is the ith component of the vector $\nabla \underline{V}$.

It is shown in standard texts on vector calculus that for a scalar function V to be obtained uniquely from a line integral of a vector function $\nabla \underline{V}$, the following $\frac{n(n-1)}{2}$ equations must be satisfied

$$\frac{\partial \nabla V_i}{\partial x_j} = \frac{\partial \nabla V_j}{\partial x_i} \qquad (i,j=1,2\ldots n) \tag{2.17}$$

These are also necessary and sufficient conditions so that the scalar function V be independent of the path of integration. In the three-dimensional case, Eqs. (2.17) are identical to those obtained by setting the curl of the vector $\nabla \underline{V}$ equal to zero.

The procedure of constructing $V(\underline{x})$ is to determine a gradient $\nabla \underline{V}$ such that

i) The n dimensional curl of $\nabla \underline{V}$ is zero.

ii) V and $\dot{V}$ is determined from $\nabla \underline{V}$ such that $V > 0$ and $\dot{V} < 0$. If $\dot{V} \leq 0$ then we must ensure that it is not zero along any solution other than the origin. If such a $\nabla \underline{V}$ is determined, then the origin is a.s.i.l. The constructive procedure consists in assuming $\nabla \underline{V}$ of the form

$$\nabla \underline{V} = \begin{bmatrix} \alpha_{11}(\underline{x}) & \alpha_{12}(\underline{x}) & \cdots & \alpha_{1n}(x) \\ \vdots & & & \\ \alpha_{n1}(\underline{x}) & \alpha_{n2}(\underline{x}) & \cdots & \alpha_{nn}(\underline{x}) \end{bmatrix} \begin{bmatrix} x_1 \\ \vdots \\ x_n \end{bmatrix} \tag{2.18}$$

α_{ij}'s are assume to consist of a constant term α_{ijk} plus a variable term α_{ijv}. For V to be positive definite in the neighborhood of the origin, the diagonal terms $\alpha_{ii}(\underline{x})$ are assumed to be of the form

$$\alpha_{ii}(\underline{x}) = \alpha_{ii}(x_i) = \alpha_{iik} + \alpha_{iiv}(x_i) \tag{2.19}$$

with $\alpha_{iik} > 0$ and $\alpha_{iiv}(x_i)$ being an even function of x_i. In this manner after integration, the α_{ii} coefficient will give rise to terms such as

$$\alpha_{iik} \frac{x_i^2}{2} \quad \text{and} \quad \int_0^{x_i} \alpha_{iiv}(u_i) u_i \, du_i$$

where u_i is a dummy variable of integration.

Further knowledge of the unknown coefficients in $\nabla\underline{V}$ is obtainable from an examination of the generalized curl equations

$$\frac{\partial \nabla V_i}{\partial x_j} = \frac{\partial \alpha_{iiv}}{\partial x_j}(\underline{x})x_1 + \cdots \frac{\partial}{\partial x_j}(\alpha_{ijk}x_j + \alpha_{ijv}(\underline{x})x_j)$$

$$\cdots \frac{\partial \alpha_{inv}}{\partial x_j}(\underline{x})x_n$$

$$\frac{\partial \nabla V_j}{\partial x_i} = \frac{\partial \alpha_{jiv}}{\partial x_i}(\underline{x})x_1 + \cdots \frac{\partial}{\partial x_i}(\alpha_{jik}x_i + \alpha_{jiv}(\underline{x})x_i)$$

$$\cdots \frac{\partial \alpha_{jnv}}{\partial x_i}(\underline{x})x_n \tag{2.20}$$

From (2.17), the right-hand sides of (2.20) are equal. Since constant terms on either side must be equal, it follows

$$\alpha_{ijk} = \alpha_{jik} \tag{2.21}$$

The other terms in $\underline{\nabla V}$ are determined jointly from the curl equations (2.17) and dV/dt. Since $dV/dt \leq 0$, an attempt is made to make it negative semidefinite in as simple a way as possible. This may be accomplished if indefinite terms in dV/dt are eliminated. In the simplest case, we assume

$$\frac{dV}{dt} = -kx_1^2 \tag{2.22}$$

where k is initially assumed a constant. If dV/dt is constrained as in the above equation, the remaining terms in dV/dt must be forced to cancel. This is accomplished by grouping terms of similar state variables and choosing α_{ij}'s to force cancellation. The α_{ij}'s are assumed constants unless cancellation or the generalized curl equations require a more complicated form. Grouping of terms is governed by the restrictions on the α_{ij}'s as stated above. For example, if in a third order system dV/dt contains the terms $\alpha_{11}x_1x_2$, $\alpha_{12}x_2^2$ and $-x_1x_2^3$, the indefinite term $-x_1x_2^3$ could not be grouped with $\alpha_{11}x_1x_2$ since α_{11} can only be a function of x_1. However, if $-x_1x_2^3$ were grouped with $\alpha_{12}x_2^2$, it could be eliminated by letting $\alpha_{12} = x_1x_2$. The choice of the α_{ij}'s to force cancellation is not arbitrary because the generalized curl equations (2.17) must also be satisfied. If dV/dt cannot be made semi-definite in one state variable, then one may try to make it semi-definite in two or three variables. Another possibility is to make dV/dt negative semi-definite in as large a region as possible around the origin in which case, we can try to find an estimate of the region of asymptotic stability by fitting the largest possible V curve using Theorem 2.7.5.

A formal five-step procedure for the application of the variable gradient method is given below:

1) Assume a gradient of the form (2.18). The α_{ij}'s may be written as though they are constants until the need arises to allow them to be more complicated.

2) From the variable gradient, form $dV/dt = \langle \underline{\nabla V}, \underline{f}(\underline{x}) \rangle$.

3) In conjunction with and subject to the requirements of the generalized curl equations (2.17), constrain dV/dt to be at least negative semi-definite.

4) From the known gradient, determine V.

5) Use the necessary theorem to establish stability, asymptotic stability in the large, or region of a.s.i.l.

In the above procedure, the most difficult step is step 3. In fact, step 3 can be conceptually broken down into smaller steps as

3(a) Constrain dV/dt to be negative semi-definite. This will give some contraint on the coefficients.

(b) Use the curl equations to determine the remaining unknown coefficients.

(c) Recheck $\dot{V}$ because the addition of terms required as a result of 3(b) may alter $\dot{V}$.

Example:

Consider the system of equations

$$\dot{x}_1 = -3x_2 - g(x_1)x_1$$

$$\dot{x}_2 = -2x_2 + g(x_1)x_1$$

The procedure outlined above will be carried out step by step.

Step 1:

The gradient function is assumed to be

$$\nabla \underline{V} = \begin{bmatrix} \alpha_{11}x_1 + \alpha_{12}x_2 \\ \alpha_{21}x_1 + \alpha_{22}x_2 \end{bmatrix}$$

Step 2:

dV/dt is determined as

$$\frac{dV}{dt} = -x_1^2[\alpha_{11}g(x_1) - \alpha_{21}g(x_1)]$$
$$+ x_1x_2[-3\alpha_{11} - \alpha_{12}g(x_1) - 2\alpha_{21} + \alpha_{22}g(x_1)]$$
$$- x_2^2(2\alpha_{22} + 3\alpha_{12})$$

Step 3:

The simplest manner in which dV/dt may be constrained to be negative semi-definite is if $\alpha_{12} = \alpha_{21} = 0$. Then

$$\frac{dV}{dt} = -\alpha_{11}g(x_1)x_1^2 - 2\alpha_{22}x_2^2 + x_1x_2(-3\alpha_{11} + \alpha_{22}g(x_1))$$

The indefinite terms in x_1x_2 may be completely eliminated if

$$\alpha_{11} = \frac{\alpha_{22}}{3} g(x_1)$$

Thus, the gradient is known to be

$$\nabla \underline{V} = \begin{bmatrix} \frac{\alpha_{22}}{3} g(x_1)x_1 \\ \alpha_{22}x_2 \end{bmatrix}$$

and $$\frac{dV}{dt} = \frac{-\alpha_{22}}{3} g^2(x_1)x_1^2 - 2\alpha_{22}x_2^2$$

Step 4:

The gradient $\underline{\nabla V}$ is integrated to yield

$$V(\underline{x}) = \int_0^{x_1} \frac{\alpha_{22}}{3} g(\gamma_1)\gamma_1 d\gamma_1 + \int_0^{x_2} \alpha_{22}\gamma_2 d\gamma_2$$

With $\alpha_{22} = 6$ as a choice of scale factors to eliminate fractions

$$V(\underline{x}) = 2\int_0^{x_1} g(\gamma_1)\gamma_1 d\gamma_1 + 3x_2^2$$

As long as $x_1 g(x_1)$ lies in the first and third quadrant, $V(\underline{x})$ is positive definite. If the integral goes to ∞ as $||x_1|| \to \infty$, then $V(\underline{x})$ is not only positive definite, but also $V(\underline{x}) = C$ represents closed surfaces for all C in the whole state space. dV/dt is negative definite in the entire state space.

Step 5:

On the basis of Theorem 2.7.3 on a.s.i.l., the system is globally asymptotically stable.

In these steps, particularly in step 3, no apparent use was made of the generalized curl equation. However, they were implicitly used, since α_{12} was a constant equal to zero and on the basis of the curl equations α_{21} was also chosen as zero. Also, since ∇V_1 does not contain x_2 and ∇V_2 does not contain x_2,

$$\frac{\partial \nabla V_1}{\partial x_2} = \frac{\partial \nabla V_2}{\partial x_1}$$

2.14 Zubov's Method [12,13]

This method enables us not only to generate a Lyapunov function, but also construct a region of attraction or an approximation to it. The crux of the method lies in solving a linear partial differential equation (p.d.e.) and when the solution to this p.d.e. is obtained in a closed form, we obtain a unique Lyapunov function and an exact stability region. When the solution is not possible in closed form, a series solution is possible and we get an approximation to the exact stability region. In particular, this method is amenable to computer solution. We first state Zubov's main theorem.

Theorem: Let U be a set containing the origin. Necessary and sufficient conditions for U to be exact domain of attraction is that two functions $V(\underline{x})$ and $\theta(\underline{x})$ exist with the following properties:

a) $V(\underline{x})$ is defined and continuous in U. $\theta(\underline{x})$ is defined and continuous in the entire state space.

b) $\theta(\underline{x})$ is positive definite for all $\underline{x}$.

c) $V(\underline{x})$ is positive definite in U with $V(\underline{0}) = 0$. Also, the inequality $0 < V(\underline{x}) < 1$ holds in U.

d) On the boundary of U, $V(\underline{x}) = 1$.

e) The following partial differential equation is satisfied

$$\sum_{i=1}^{n} \frac{\partial V}{\partial x_i} f_i(\underline{x}) = -\theta(x)(1-V(\underline{x}))(1+\|\underline{f}\|^2) \qquad (2.23)$$

<u>Comment:</u> In the above theorem, if the domain of attraction is the entire state space, we then have global asymptotic stability and the condition on $V(\underline{x})$ is then

$$V(\underline{x}) \rightarrow 1 \quad \text{as} \quad ||\underline{x}|| \rightarrow \infty$$

<u>Alternative Formulations of the p.d.e. (2.23):</u>

1. Since the term $(1+||\underline{f}||^2)$ is positive, we may also define another positive definite function $\phi(\underline{x})$ as

$$\phi(\underline{x}) = \theta(\underline{x})(1+||\underline{f}||^2)$$

The p.d.e. then becomes

$$\sum_{i=1}^{n} \frac{\partial V}{\partial x_i} f_i(\underline{x}) = -\phi(\underline{x})(1-V(\underline{x})) \tag{2.24}$$

2. In another formulation of Zubov's method, the right hand side of the p.d.e. (2.23) appears as

$$-\theta(\underline{x})(1-V(\underline{x}))(1+||\underline{f}||^2)^{1/2} \tag{2.25}$$

3. If we define a change of variable $\hat{V} = -\, n(1-V)$, the p.d.e. (2.24) becomes

$$\sum_{i=1}^{n} \frac{\partial \hat{V}}{\partial x_i} f_i(\underline{x}) = -\phi(\underline{x}) \tag{2.26}$$

and the Lyapunov function is $\hat{V}$. The region of attraction then becomes $0 < \hat{V} < \infty$. This form of p.d.e. is sometimes helpful in obtaining $V(\underline{x})$ in a series form which converges rapidly.

<u>Example:</u>

$$\dot{x}_1 = -x_1 + x_2 + x_1(x_1^2+x_2^2)$$

$$\dot{x}_2 = -x_1 + x_2 + x_2(x_1^2+x_2^2)$$

Assume a function $\phi(x) = 2(x_1^2+x_2^2)$. Substituting in (2.24),

$$\frac{\partial V}{\partial x_1}(-x_1+x_2+x_1^3+x_1x_2^2) + \frac{\partial V}{\partial x_2}(-x_1-x_2+x_2^3+x_1^2x_2)$$

$$= -2(x_1^2+x_2^2)(1-V(\underline{x}))$$

It turns out that $V(\underline{x}) = x_1^2 + x_2^2$ satisfies the p.d.e. Hence, the required Lyapunov function is $V(\underline{x}) = x_1^2 + x_2^2$ and the stability boundary is given by

$$x_1^2 + x_2^2 = 1$$

Example:

Consider the system

$$\dot{x}_1 = -x_1 + 2x_1^2 x_2$$

$$\dot{x}_2 = -x_2$$

Using the formulation in (2.25), the partial differential equation is

$$\frac{\partial V}{\partial x_1}(2x_1^2 x_2 - x_1) + \frac{\partial V}{\partial x_2}(-x_2)$$

$$= -\theta(\underline{x})\{(1-V)[1 + x_2^2 + (2x_1^2 x_2 - x_1)^2]^{1/2}\}$$

Let $\theta(\underline{x}) = \dfrac{x_1^2 + x_2^2}{1 + x_2^2 + (2x_1^2 x_2 - x_1)^2}$

Clearly $\theta(x_1, x_2)$ satisfies the condition of the theorem. The solution to the p.d.e. is given by

$$V = 1 + \exp\left(\frac{-x_2^2}{2} - \frac{-x_1^2}{2(1-x_1 x_2)}\right)$$

$V(x_1, x_2)$ vanishes at the origin and it is equal to 1 on the curve $x_1 x_2 = 1$. Hence, the region of stability is defined by the interior of the curve $x_1 x_2 = 1$.

2.14.1 Series Solution for Zubov's Method

In general, we cannot expect a closed form solution to the partial differential equation in Zubov's method. If the non-linearity is expressible in an analytic manner, e.g. power series, then Zubov's method can be used to construct the Lyapunov function as well as approximate the region of attraction.

Let $\underline{\dot{x}} = \underline{f}(\underline{x})$ be expanded as

$$\underline{\dot{x}} = \underline{A}\,\underline{x} + \underline{g}(\underline{x}) \tag{2.27}$$

where $\underline{g}(\underline{x})$ contains terms of second degree and higher and $\underline{A}$ is the linear part. Assume $\underline{A}$ to be stable i.e. $\underline{A}$ has all eigenvalues with negative real parts. We choose $\phi(\underline{x})$ to be a positive definite quadratic form. The solution of the p.d.e.

$$\sum_{i=1}^{n} \frac{\partial V}{\partial x_i} f_i(\underline{x}) = -\phi(\underline{x})(1-V(\underline{x})) \tag{2.28}$$

is sought in the form

$$V(\underline{x}) = V_2(\underline{x}) + V_3(\underline{x}) + \ldots . \tag{2.29}$$

where $V_2(\underline{x})$ is quadratic in $\underline{x}$ and $V_m(\underline{x})(m=3,4,\ldots)$ are homogeneous in degree m, i.e. $V_m(\gamma\,\underline{x}) = \gamma^m V(\underline{x})$ for any constant γ. To find $V_m(\underline{x})(m=2,3,\ldots)$ we substitute the system differential equations (2.27) and $V(\underline{x})$ given by (2.29) in (2.28). Because of the assumptions on $\underline{g}(\underline{x})$ and $V_m(\underline{x})$, $V_2(\underline{x})$ is found as the Lyapunov function for the linear equation

$$\underline{\dot{x}} = \underline{A}\,\underline{x} \tag{2.30}$$

$V_m(\underline{x})$, $(m=3,4,\ldots)$ are found from the recurrence relations

$$\sum_{i=1}^{n} \frac{\partial V_m}{\partial x_i} f_i(\underline{x}) = R_m(\underline{x}) \tag{2.31}$$

which simplifies to

$$\left[\frac{\partial V_m}{\partial x_1} \cdots \frac{\partial V_m}{\partial x_n}\right] [\underline{A}\,\underline{x}] = R_m(\underline{x}) \quad m=3,4,\ldots \tag{2.32}$$

$R_m(\underline{x})$ will be known if $V_2(\underline{x})\ldots V_{m-1}(\underline{x})$ have already been determined. Thus, $V(\underline{x})$ is found as $\Sigma V_i(\underline{x})$ and the number of terms to be taken in $V(\underline{x})$ depends upon the particular problem under consideration. The region of asymptotic stability obtained by taking more terms does not necessarily converge to the exact stability boundary uniformly. It is possible to substitute $\phi(\underline{x})$ in (2.28) by means of a positive semi-definite quadratic function instead of a positive definite quadratic function. Because of the assumption on $\underline{A}$ we are still guaranteed to get a positive definite $V_2(\underline{x})$.

2.14.2 Computation of Region of Attraction

The procedure is outlined below with the computer implementation in mind for computing the region of asymptotic stability. Compute $\dot{V}_2(\underline{x})$ with $\dot{\underline{x}} = \underline{A}\,\underline{x} + \underline{g}(\underline{x})$. Since $\underline{A}$ is stable, $\dot{V}_2 \leq 0$. Let W_2 be the set of points on which $\dot{V}_2(\underline{x}) = 0$. Let $\alpha = \text{Min}\ V_2(\underline{x})$ and $\beta = \text{Max}\ V_2(\underline{x})$ for $\underline{x} \varepsilon W_2$. According to Zubov, the curve defined by $V_2(\underline{x}) = \alpha$ lies entirely inside the region of attraction of the equilibrium point and the curve defined by $V_2(\underline{x}) = \beta$ lies entirely outside the region of attraction. These two assertions imply that the boundary of the region of asymptotic stability lies in the region defined by

$$\alpha < V_2(\underline{x}) < \beta$$

We can extend the above reasoning to $V_2 + V_3$, $V_2 + V_3 + V_4, \ldots$, etc. To generalize, define

$$V^{(n)}(\underline{x}) = V_2(\underline{x}) + V_3(\underline{x}) + \ldots V_n(\underline{x})$$

Let $W^{(n)}(\underline{x})$ = set of all points on which $\dot{V}^{(n)}(\underline{x}) = 0$. Specifically, we ensure that $W^{(n)}(\underline{x})$ comprises all points of zero $\dot{V}^{(n)}(\underline{x})$ which define boundaries between regions of positive and negative $\dot{V}^{(n)}(\underline{x})$. Let $\alpha^{(n)}$ be the smallest value of $V^{(n)}$ on $W^{(n)}$ and $\beta^{(n)}$ be the largest value of $V^{(n)}$ on $W^{(n)}$. Let $A^{(n)}$ be the set of points $\underline{x}$ for which $V^{(n)}(\underline{x}) \leq \alpha^{(n)}$. We can then say that the curve $V^{(n)}(\underline{x}) = \alpha^{(n)}$ is wholly contained in the region of attraction and the curve $V^{(n)}(\underline{x}) = \beta^{(n)}$ wholly lies outside the region of attraction.

In implementing the procedure for a numerical problem, we successively compute $\alpha^{(n)}$ and the region $A^{(n)}$. One would be tempted to infer that as n takes on large values, $A^{(n)}$

approximates better and better to the exact stability region. However, this is not necessarily true, i.e.

$$A^{(n)} \not\subset A^{(n+1)}$$

2.14.3 Analytical Details of Series Solution

Consider the system of differential equations (2.27) written in the summation form as

$$\frac{dx_i}{dt} = \sum_{j=1}^{n} a_{ij}x_j + \sum_{\ell \geq 2} p_i(m_1,\ldots,m_n)x_1^{m_1} x_2^{m_2} \ldots x_n^{m_n}$$

$$\ell = m_1 + m_2 \ldots m_n \qquad (i=1,2,\ldots,n) \tag{2.33}$$

where the a_{ij} and $p_i(m_1,m_2,\ldots,m_n)$ are real numbers. Also assume ϕ of the form $\phi_2 + \phi_3 + \ldots$ where ϕ_i is a term homogeneous of degree i. Then for $V_2(\underline{x})$ we have

$$\sum_{i=1}^{n} \frac{\partial V_2}{\partial x_i} \sum_{j=1}^{n} a_{ij}x_j = -\phi_2(\underline{x}) \tag{2.34}$$

and the other recurrence relations corresponding to (2.32) become

$$\sum_{i=1}^{n} \frac{\partial V_m}{\partial x_i} \sum_{j=1}^{n} a_{ij}x_j = R_m(\underline{x}) \qquad m=3,4,\ldots. \tag{2.35}$$

where $R_m(\underline{x})$ is formed from ϕ_m and the already known V_i's.

If we compare like terms on both sides in (2.34) and (2.35), then a set of linear algebraic equations result which may be solved for coefficients of $V_m(m=2,3,\ldots)$. The series is generally truncated up to a certain value of m.

2.14.3 Computation Example

This example is adapted from Ref. [12]. Consider the modified form of the well known Vanderpol's equation

$$\ddot{x} + \varepsilon(1-x^2)\dot{x} + x = 0 \tag{2.36}$$

This is obtained from the usual Vanderpol's equation by putting t = -t so that we have a region of attraction around the origin. Introduce the variables defined by

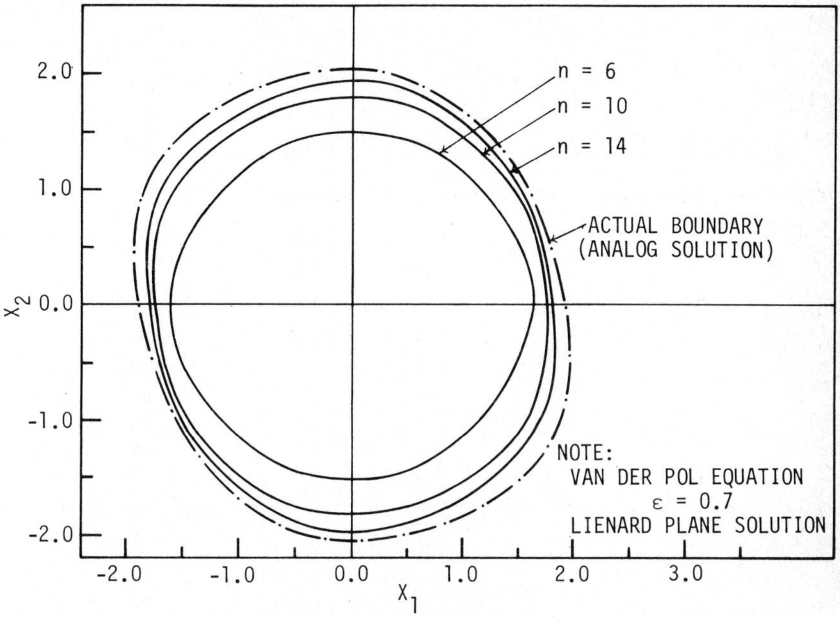

Fig. 2.5. Stability boundary as given by approximate solutions to Zubov equation. (Reproduced from Ref. [12].

$$\dot{x}_1 = x$$
$$\dot{x}_2 = \dot{x} + \quad (x - \frac{x^3}{3}) \tag{2.37}$$

and thus obtain

$$\dot{x}_1 = x_2 - \varepsilon(x_1 - \frac{x_1^3}{3})$$
$$\dot{x}_2 = -x_1 \tag{2.38}$$

In the state space description (2.38), we have the state trajectories which plot x against its (negative) time integral. This contrasts with the usual procedure of x versus its time derivative. The results can be transferred to the usual phase plane by solving for $\dot{x}$ from (2.37).

Assume $V(x_1,x_2)$ of the form

$$V(x_1,x_2) = \sum_{j=2}^{n} \sum_{k=0}^{j} d_{jk} \, x_1^{\,j-k} \, x_2^{\,k} \tag{2.39}$$

The terms $V_2(x_1,x_2)$, $V_3(x_1,x_2)$.... are computed recursively using the procedure outlined in the preceding section. The procedure will be illustrated for n=3.

$$\begin{aligned} V^{(3)} &= V_2 + V_3 \\ &= (d_{20}x_1^2+d_{21}x_1x_2+d_{22}x_2^2) \\ &+ (d_{30}x_1^3+d_{31}x_1^2x_2+d_{32}x_1x_2^2+d_{33}x_2^3) \end{aligned} \tag{2.40}$$

Choose $\phi(x_1,x_2) = x_1^2 + x_2^2$. The p.d.e. (2.35) with $V(\underline{x}) = V^{(3)}$ now becomes

$$\begin{aligned} &\frac{\partial V^{(3)}}{\partial x_1} \left(x_2-\varepsilon\left(x_1 - \frac{x_1^3}{3}\right)\right) + \frac{\partial V^{(3)}}{\partial x_2} (-x_1) \\ &= -(x_1^2+x_2^2)(1-V^{(3)}(\underline{x})) \end{aligned} \tag{2.41}$$

Equating terms of same degree on both sides, we get

$$\begin{aligned} &\frac{\partial V_2}{\partial x_1} (x_2-\varepsilon x_1) + \frac{\partial V_2}{\partial x_2} (-x_1) = -(x_1^2+x_2^2) \\ &\frac{\partial V_3}{\partial x_1} (x_2-\varepsilon x_1) + \frac{\partial V_3}{\partial x_2} (-x_1) = 0 \end{aligned} \tag{2.42}$$

The appearance of 0 term on the right side of the second equation in (2.42) for terms of degree 3 is not surprising because there is no second degree term in the nonlinearity and, furthermore, $\phi(\underline{x})$ has been chosen as $\phi_2(\underline{x})$. In Ref. [12], computations have been done by taking $\phi_2(\underline{x}) = x_1^2$ and region of attraction computed (Fig. 2.5) as discussed in Sec. 2.14.2. ε is taken as 0.7. For n=2, the stability region is found to be a circle of radius $\sqrt{3}$. For n=6, the stability boundary is not improved, but contained in the curve for n=2. On the other hand, n=10 approximation encloses both the n=2 and n=6 approximation and is in turn completely enclosed by the n=14 approximation. Thus, a surprisingly high order approximation must be used in order to improve upon the quadratic results. References [12] and [12] contain a good discussion regarding the numerical aspects of the solution.

2.15 Absolute Stability - Popov's Method

So far, we have been considering general nonlinear systems without any restrictions on the nonlinearities. We also found that for such systems there is no systematic method to construct Lyapunov functions in order to ascertain stability. In addition to the classical method based on first integrals of motion and Krasavoskii's method, two other popular methods were discussed, namely, variable gradient and Zubov's method. However, if the nonlinearity is such that it lies in the first and third quadrant or in a sector thereof, then a systematic procedure to construct the Lyapunov function is possible. The problem therefore is to investigate the stability of such systems. In 1944, Luré [14] constructed a Lyapunov function for such systems consisting of the sum of a quadratic form in the state variables and an integral of the nonlinearity and derived therefrom a set of nonlinear equations called Lure's resolving equations. The existence of a set of solutions for such equations constituted sufficient conditions for global asymptotic stability of the nonlinear system. A significant breakthrough in this area of investigation came from the contribution of V. M Popov in 1962 [2]. Popov outlined a frequency domain criterion for the stability of a nonlinear system containing a single nonlinearity lying in the first and third quadrant. Kalman [15] and Yakubovitch [16] established the connection between Popov's frequency domain criterion and the Lyapunov function of the type of quadratic form plus integral of the nonlinearity by proving that satisfaction of the former is a necessary and sufficient condition for the existence of the latter. Subsequently, a number of research workers have extended the results to nonlinear systems with multiple nonlinearities as well as discrete time systems, etc. [17]. By now, the literature on the topic of absolute stability is vast and scattered both in the mathematical and the engineering literature.

2.15.1 Single Nonlinearity Case

Consider the system in state space form

$$\begin{aligned} \dot{\underline{x}} &= \underline{A}\,\underline{x} + \underline{b}\,\xi \\ \xi &= -\phi(\sigma) \\ \sigma &= \underline{c}^T\underline{x} \end{aligned} \tag{2.43}$$

where $\underline{A}$ is a n×n stable matrix, $\underline{x}$, $\underline{b}$, $\underline{c}$ are n-vectors. $\phi(\sigma)$ is a nonlinearity which lies in the first and third quadrant. The system (2.43) can be cast in the block diagram as in Fig. 2.6 which consists of a linear transfer function with a nonlinearity in the feedback path. It can be verified from (2.43) that the transfer function of the linear part is

$$G(s) = \frac{\sigma(s)}{\xi(s)} = \underline{c}^T(s\underline{I}-\underline{A})^{-1}\underline{b} \tag{2.44}$$

If $\underline{A}$ is a stable matrix, i.e. all eigenvalues are in the left half plane, then G(s) has all poles with negative real parts.

Special cases of interest are when G(s) has poles on the imaginary axis. We will consider only the simplest of these cases, namely, when G(s) has a single pole at the origin. In terms of the state space description, this means that the $\underline{A}$ matrix has a zero eigenvalue. The state space description then is of the form

$$\begin{bmatrix} \dot{\underline{x}} \\ \dot{\xi} \end{bmatrix} = \begin{bmatrix} \underline{A} & \underline{0} \\ \underline{0} & \underline{0} \end{bmatrix} \begin{bmatrix} \underline{x} \\ \xi \end{bmatrix} + \begin{bmatrix} \underline{b} \\ 1 \end{bmatrix} u \tag{2.45}$$

$$u = -\phi(\sigma)$$

$$\sigma = \underline{c}^T\underline{x} + d\,\xi$$

In addition to $\underline{A}$ being a Hurwitz matrix in both (2.43) and (2.45), we will further assume that (i) the pair $(\underline{A},\underline{b})$ is

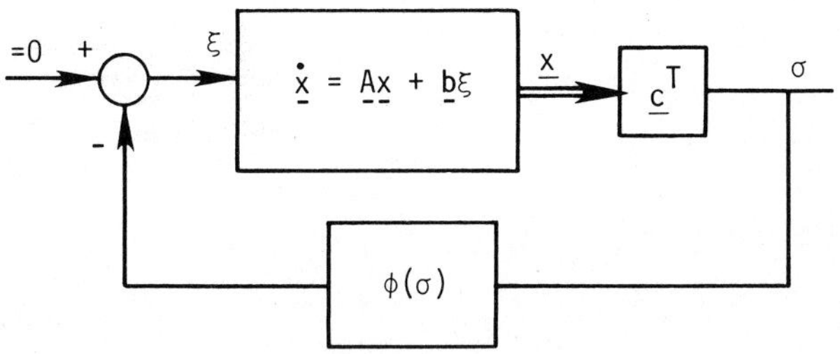

Fig. 2.6. Block diagram for the system (2.43).

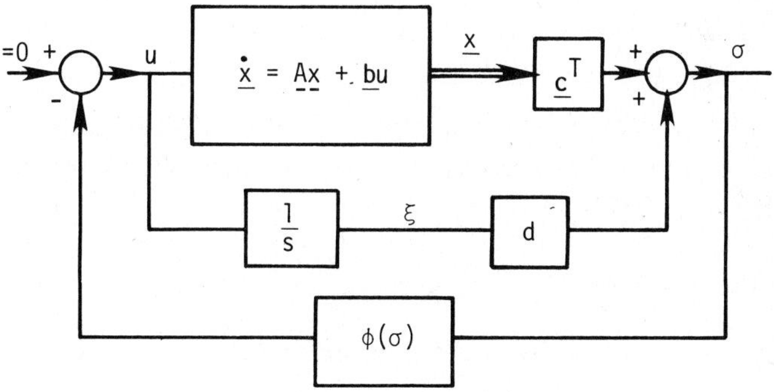

Fig. 2.7. Block diagram for the system (2.45).

controllable, and (ii) $(\underline{A}, \underline{c}^T)$ is observable. The nonlinearity satisfies the sector condition, i.e.

$$0 \leq \phi(\sigma) \leq k\sigma^2 \tag{2.47}$$

for the system (2.43) and

$$0 < \phi(\sigma) < k\sigma^2 \tag{2.48}$$

for the system (2.45).

2.15.2 Popov's Theorem

The system (2.43) or (2.45) is absolutely stable, i.e., the origin is a.s.i.l. for all nonlinearities satisfying the sector condition (2.47) or (2.48) respectively if there exists a finite real q such that

$$\frac{1}{k} + \mathrm{Re}\{(1+j\omega q)G(j\omega)\} > 0 \quad \text{for all} \quad \omega \geq 0. \tag{2.49}$$

In the case of the system (2.45), an additional necessary condition, namely, $d > 0$ is to be satisfied. It is noted from the transfer function that d is the residue of G(s) at $s = 0$. This condition is physically equivalent to the condition that the system (2.45) is stable if $\phi(\sigma) = \varepsilon\sigma$ $(\varepsilon > 0$ and small).

2.15.3 Geometric Interpretation

For the single nonlinearity case, a simple geometric interpretation can be given to the Popov criterion. Let

$$G(j\omega) = X(\omega) + j\,Y(\omega)$$

Then (2.49) becomes

$$\frac{1}{k} + \mathrm{Re}[1+j\omega q][X(\omega) + j\,Y(\omega)] > 0$$

i.e.

$$\frac{1}{k} + X(\omega) - q\omega\,Y(\omega) > 0 \tag{2.50}$$

Let

$$X^*(\omega) = X(\omega) \quad \text{and} \quad Y^*(\omega) = \omega\,Y(\omega)$$

Then we can define modified frequency response $G^*(j\omega)$ as

$$G^*(\omega) = X^*(\omega) + j\, Y^*(\omega)$$

Expression (2.50) becomes

$$\frac{1}{k} + X^*(\omega) - q\, Y^*(\omega) > 0 \tag{2.51}$$

Geometrically, we can interpret (2.51) in the modified frequency response plane as implying that there must exist a line called the Popov line passing through the point $(-\frac{1}{k}, 0)$ having a slope $\frac{1}{q}$ so that $G^*(j\omega)$ lies to the right of this line (Fig. 2.8). Thus, absolute stability, i.e., a.s.i.l. of the origin for nonlinearities satisfying the sector condition (2.47) or (2.48) can be ascertained in a simple manner analogous to the Nyquist's criterion for linear systems. An additional benefit of Popov's criterion is that if we can establish absolute stability, then we also have a systematic procedure to construct a Lyapunov function. This Lyapunov function is of the quadratic plus integral of the nonlinearity type. Systematic construction of the Lyapunov function is possible through the solution of a set of nonlinear algebraic equations (2.52) and (2.53) which are written for the infinite sector case, i.e. $k = \infty$. These equations form part of the Kalman-Yakubovitch Lemma [15,16] which is discussed in detail in Refs. [6,17,19].

<u>For the System (2.43)</u>

$$\underline{A}^T\underline{P} + \underline{P}\underline{A} = -\varepsilon\, \underline{Q} - \underline{u}\, \underline{u}^T$$

$$\underline{P}\, \underline{b} = \frac{1}{2}\beta\, \underline{A}^T\underline{c} + \alpha\, \underline{c} + (\beta\, \underline{c}^T\underline{b})^{1/2}\, \underline{u} \tag{2.52}$$

$q = \frac{\beta}{\alpha}$, $\underline{u}$ is n-vector, $\varepsilon \geq 0$ and small.

<u>For the System (2.45)</u>

$$\underline{A}^T\underline{P} + \underline{P}\, \underline{A} = -\varepsilon\, \underline{Q} - \underline{u}\, \underline{u}^T$$

$$\underline{P}\, \underline{b} = \frac{1}{2}\beta\, \underline{A}^T\underline{c} + \alpha\, d\underline{c} + [\beta(\underline{c}^T\underline{b}+d)]^{1/2}\, \underline{u} \tag{2.53}$$

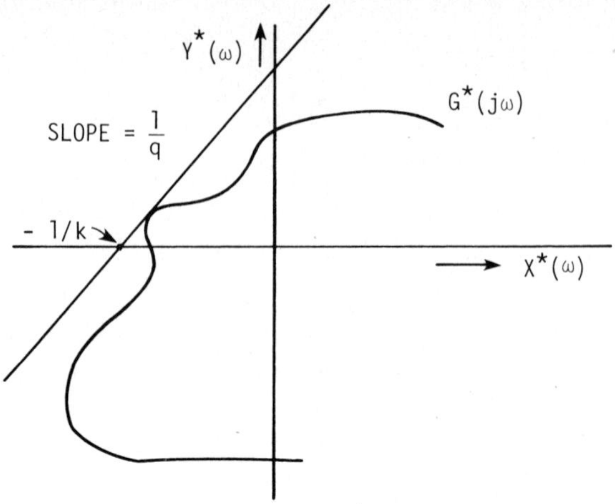

Fig. 2.8. Polar plot of the modified frequency response and the Popov line.

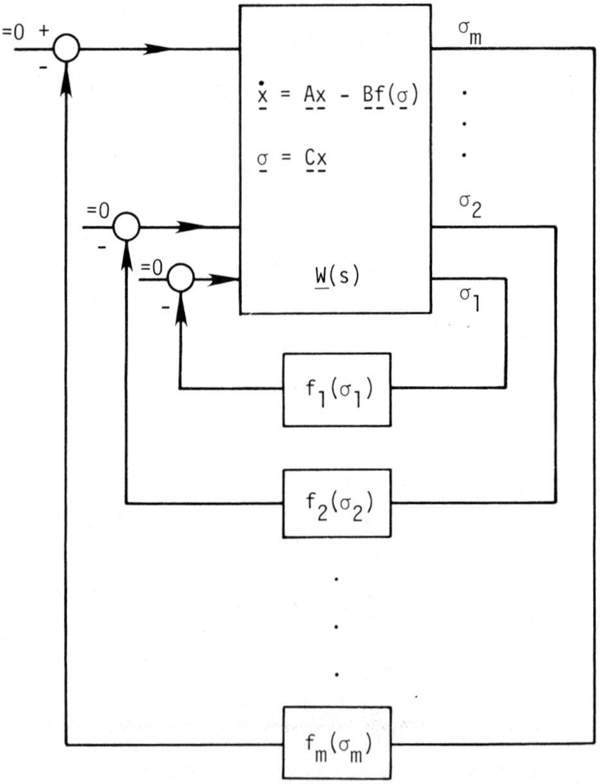

Fig. 2.9. Multivariable nonlinear system.

where $q = \frac{\beta}{2\alpha d}$, ε is ≥ 0 and small and $\underline{u}$ is (n-1) vector.

Although the form of (2.52) and (2.53) appear similar, the former is not the special case of the latter.

If Popov's criterion is satisfied, we know q. Then by an arbitrary choice of α, β and $\underline{u}$, a p.d. solution for $\underline{P}$ can be found from (2.52) or (2.53) and the Lyapunov function is then of the form

For the system (2.43)

$$V(\underline{x}) = \underline{x}^T \underline{P}\, \underline{x} + q \int_0^\sigma \phi(\sigma) d\sigma \tag{2.54}$$

For the system (2.45)

$$V(\underline{x}) = \underline{x}^T \underline{P}\, \underline{x} + \frac{1}{2} d\xi^2 + q \int_0^\sigma \phi(\sigma) d\sigma \tag{2.55}$$

In arriving at (2.54) and (2.55) the choice of $\alpha = 1$ and $\alpha = \frac{1}{2d}$ has been made respectively so that $q = \beta$.

2.15.4 Kalman's Construction Procedure

The solution of nonlinear algebraic equations can be circumvented for the single nonlinearity case by what is known as the Kalman's construction procedure [15].

The various steps in the procedure are:

1. Find q from the Popov Criterion.
2. Write down the following polynomial of order 2(n-1) in ω.
   $$W(\omega) = \mathrm{Re}[(1+jq\omega)G(j\omega)\psi(j\omega)\psi(-j\omega)] \tag{2.56}$$
   where $\psi(s) = \det(s\underline{I}-\underline{A})$.
3. Factorize $W(\omega) = \theta(j\omega)\ \theta(-j\omega)$ (2.57)
4. Let $j\omega = z$. Then $\theta(z)$ will be a polynomial in z of formal order n for the system (2.43) and (n-1) for system (2.45) with leading coefficient $\sqrt{r}$ where $r = q\, \underline{c}^T\underline{b}$ for (2.43) and $q(d+\underline{c}^T\underline{b})$ for (2.45).
5. Define $\nu(z) = -\theta(z) + \sqrt{r}\ \psi(z)$ and arrange this (2.58) polynomial of formal order 2n for the system (2.43) (2n-2) for the system (2.45) in ascending powers of z.

6. Define a real n-vector $\underline{u}$ for the system (2.43) or (n-1)-vector for the system (2.45) with its components being the coefficients of the above $\nu(z)$. Solve for $\underline{P}$ using the Lyapunov matrix equation

$$\underline{A}^T\underline{P} + \underline{P}\ \underline{A} = -\underline{u}\ \underline{u}^T \tag{2.59}$$

The Lyapunov function is then given by either (2.54) or (2.55).

2.16 Multivariable Popov Criterion

There are several versions extending Popov's criterion to systems containing multiple nonlinearities. Perhaps the most popular among these is using the Moore-Anderson Theorem [18].

Consider the system in Fig. 2.9 where $\underline{W}(s)$ is an m×m matrix transfer function such that $\underline{W}(\infty) = 0$. The nonlinearity $\underline{f}(\underline{\sigma})$ of dimension m is assumed to satisfy the properties

i) $f_k(\underline{\sigma}) = f_k(\sigma_k)$

ii) $f_k(\sigma_k)\sigma_k > 0 \qquad k=1,2,\ldots m$

(ii) implies that each nonlinearity lies in the first and third quadrant. Let $\underline{A}$, $\underline{B}$, $\underline{C}$ constitute the minimal realization of $\underline{W}(s)$, i.e.

$$\underline{W}(s) = \underline{C}(s\underline{I}-\underline{A})^{-1}\underline{B}$$

The state space description of the system is given in the form

$$\begin{aligned} \dot{\underline{x}} &= \underline{A}\ \underline{x} - \underline{B}\ \underline{f}(\underline{\sigma}) \\ \underline{\sigma} &= \underline{C}\ \underline{x} \end{aligned} \tag{2.60}$$

where $\underline{A}$, $\underline{B}$, $\underline{C}$ are n×n, n×m and m×n matrices respectively.

Moore-Anderson Theorem: If there exists real matrices $\underline{N}$ and $\underline{Q}$ such that

$$\underline{Z}(s) = (\underline{N} + \underline{Q}\ s)\underline{W}(s) \tag{2.61}$$

is positive real, then the system (2.60) is asymptotically stable in the large providing $(\underline{N} + \underline{Q}\ s)$ does not cause pole-zero cancellation with $\underline{W}(s)$.

The conditions for $\underline{Z}(s)$ to be positive real are

i) $\underline{Z}(s)$ has elements which are analytic for Re(s) > 0.

ii) $\underline{Z}^*(s) = \underline{Z}(s^*)$ for Re s > 0.

iii) $\underline{Z}^T(s^*) + \underline{Z}(s)$ is positive semi-definite for Re (s) > 0.

Conditions (i) and (ii) hold for most physical systems and condition (iii) implies

$$\underline{Z}^T(-j\omega) + \underline{Z}(j\omega) \geq 0 \quad \text{for} \quad \omega > 0. \tag{2.62}$$

Analogous to the nonlinear equations (2.52) and (2.53) for the single nonlinearity case, we have a method of constructing the Lyapunov function for (2.60) if the Moore-Anderson theorem is satisfied.

Theorem [18]

Given $\underline{N}$ and $\underline{Q}$ satisfying the multivariable Popov criterion (2.61), then a Lyapunov function of the "quadratic plus integral of nonlinearity" exists of the form

$$V(\underline{x}) = \underline{x}^T\underline{P}\,\underline{x} + \int_0^{\underline{\sigma}} \underline{f}^T(\underline{\sigma})\,\underline{Q}\,d\underline{\sigma} \tag{2.63}$$

and $\underline{P}$ is obtained as a positive definite matrix satisfying the following set of nonlinear algebraic equations

$$\begin{aligned} \underline{A}^T\underline{P} + \underline{P}\,\underline{A} &= -\underline{L}\,\underline{L}^T \\ \underline{P}\,\underline{B} &= \underline{C}^T\underline{N} + \underline{A}^T\underline{C}^T\underline{Q} - \underline{L}\,\underline{W} \\ \underline{W}^T\underline{W} &= \underline{Q}\,\underline{C}\,\underline{B} + \underline{B}^T\underline{C}^T\underline{Q} \end{aligned} \tag{2.64}$$

where $\underline{L}$ and $\underline{W}$ are auxiliary matrices.

Comments

i) Unlike in the single nonlinearity case, an easy geometric interpretation of the multivariable Popov criterion is not possible.

ii) Extensions of the above basic theorem to finite sector case are available in the literature [17].

iii) The algebraic equations are not amenable to a systematic solution procedure like Kalman's construction procedure for the single nonlinearity cases unless we resort to techniques like spectral factorization, etc. Fortunately, as we shall see in Chapter IV, for the power system case the mathematical model results in a solution of (2.64) for $\underline{P}$ easily.

iv) Although not explicitly stated, the condition on $\underline{A}$ for the theorem to hold is that it be stable. The poles on the imaginary axis if they exist must be simple.

Conclusion

In this chapter, we have discussed various methods of constructing the Lyapunov function in a systematic manner. We have studied the basic stability definition in the sense of Lyapunov and theorems which give sufficient conditions for stability of various types. Popov's criterion, both for single nonlinearity as well as multinonlinear cases, have been discussed.

References

1. A. M. Lyapunov, "Probleme Général de la Stabilité du Movement", Reprinted in Annals of Mathematical Studies No. 17, Princeton University Press, Princeton, N.J., 1949 (Russian Edition 1892).
2. V. M. Popov, "Absolute Stability of Nonlinear Systems of Automatic Control", Automation and Remote Control, Vol. 22, 1962, pp. 857-875.
3. W. Hahn, "Theory and Application of Liapunov's Direct Method", (Book) Prentice Hall, Englewood Cliffs, N.J., 1963.
4. J. P. LaSalle and S. Lefschetz, "Stability by Liapunov's Direct Method with Applications", (Book) Academic Press, New York, 1961.
5. J. L. Willems, "Stability Theory of Dynamical Systems", (Book) Thomas Nelson & Sons, U.K., 1970.
6. M. Vidyasagar, "Nonlinear Systems Analysis", (Book) Prentice Hall, Englewood Cliffs, N.J., 1978.
7. B. D. O. Anderson and S. Vongpanitlerd, "Network Analysis and Synthesis - A Modern Systems Theory Approach", (Book) Prentice Hall, Englewood Cliffs, N.J.
8. O. Gurel and L. Lapidus, "A Guide to the Generation of Lyapunov Functions", Industrial and Engineering Chemistry, March 1969, pp. 30-41.
9. N. G. Chetaev, "Stability of Motion", (Book) Pergamon Press, 1961 (Russian Edition, 1946-1950).
10. N. N. Krasovskii, "Stability of Motion", (Book) Stanford University Press, Stanford, California, 1963 (Russian Edition 1959).
11. D. G. Schultz and J. E. Gibson, "The Variable Gradient Method for Generating Liapunov Functions", AIEE Trans. Part II, Appl. & Industry, Vol. 81, 1962, pp. 203-210.
12. S. G. Margolis and W. G. Vogt, "Control Engineering Applications of V. I. Zubov's Construction Procedure for Lyapunov Functions", IEEE Trans. Automatic Control, Vol. AC-8, No. 2, April 1963, pp. 104-113.
13. Discussion on Ref. 12 by F. Fallside and M. R. Patel and Reply, IEEE Trans. Automatic Control, Vol. AC-10, No. 2, April 1965, pp. 220-222.
14. A. I. Luré and V. N. Postnikov, "On the Theory of Stability of Control Systems", Prikl. Mat. i. Mehk, Vol. 8, No. 3, 1944.
15. R. E. Kalman, "Lyapunov Functions for the Problem of Luré in Automatic Control", Proc. Nat. Acad. Sci., Vol. 49, February 1963, pp. 201-205.

16. V. A. Yakubovitch, "The Solution of Certain Matrix Inequalities in Automatic Control", Dokl. Akad. Nauk. S.S.S.R., Vol. 143, 1962, pp. 1304-1307.

17. K. S. Narendra and J. H. Taylor, "Frequency Domain Criteria for Absolute Stability", (Book) Academic Press, New York, 1973.

18. J. B. Moore and B. D. O. Anderson, "A Generalization of the Popov Criterion", Jour. Franklin Institute, Vol. 285, No. 6, June 1968, pp. 488-492.

19. M. A. Aizerman and F. R. Gantmacher, "Absolute Stability of Regulator Systems", (Book) Holden-Day, Inc., San Francisco, U.S.A., 1964. (Revision Edition 1963)

Chapter III

MATHEMATICAL MODELS OF MULTI-MACHINE POWER SYSTEMS

3.1 Introduction

As a preliminary step to the application of Lyapunov's method for power systems, we will develop the necessary mathematical model of a multi-machine power system in this chapter. The bulk of literature in Lyapunov stability analysis of power systems rests on a mathematical model involving a number of simplifying assumptions. The mathematical model can be retained in the form of a set of second order nonlinear differential equations or conveniently be cast in the Luré form with the nonlinearities in the first and third quadrant over a region around the origin. This enables the systematic construction of the Lyapunov function and also the computation of the region of attraction. Both formulations are equivalent and are prevalent in the literature. Towards the end of the chapter, we present the mathematical model relaxing some of the simplifying assumptions and which give rise to multiplicative nonlinearities in the mathematical model.

Since synchronism is essentially a physical phenomena associated with relative motion between synchronous machines, some sort of reference angle other than the synchronous reference frame angle is necessary. There are two possibilities here. We can take the angle of one of the machines as reference angle (generally the machine having the largest inertia) and measure angles of other machines with respect to this machine. The other possibility is to define a center of angle proportional to inertia weighted angles of all the machines, a concept analogous to the center of mass in mechanical systems. All machine angles are then measured with respect to this center of angle. The state space models based on both these notions are developed and compared. The correct dimension of the state space for the different models is derived from physical as well as control theoretic concepts of controllability, observability, minimal realization theory, etc. The Luré form of equations are then derived which constitute the basis for systematic contruction of Lyapunov functions and also for applying the Popov stability criterion. Both the cases of zero and non-negligible transfer conductances are considered as well as zero and non-zero damping. Finally, the mathematical model for the multi-machine case including the flux decay effect is considered.

3.2 Transient Stability Problem Restated

By transient stability, we imply the stability of the system to maintain synchronism under sudden and major impacts. A fault on a high voltage transmission line, loss of a large generating unit, are examples of large disturbances. Let us first see what happens actually on the occurrence of a major disturbance. A major disturbance basically creates an imbalance between generation and load. Consequently, the power balance at each generating unit (i.e., mechanical input power minus the electrical output power) differs considerably from generator to generator. Some of them will have net accelerating power while others have decelerating power. As a result, the rotor angles of the machines accelerate or decelerate beyond the synchronous speed for $t > 0$. This is called "swinging" of the machines. If the rotor angles are plotted as a function of time, there exist two possibilities. (See Figs. 1.1 and 1.2)

i) The rotor angle increase together and swing in unison and ultimately settle to new angles. Since the relative rotor angles do not increase, we say the system is stable and synchronism is preserved.

ii) One or more of the machine angles accelerate faster than the rest of them, so that ultimately the rotor angles of the faster group continues to increase indefinitely. The relative rotor angles diverge. Such a mode of behavior is classified as unstable or losing synchronism.

If the fault is cleared quickly, the tendency to separate may be arrested through the post-fault network by a redistribution of the energy gained by the system during the faulted period. The maximum time through which a fault can be allowed to remain without losing synchronism subsequently is called the critical clearing time t_{cr}.

All of the above concepts which are mainly physical have to be translated into mathematical ideas before Lyapunov's theory can be applied. We do this first by developing the mathematical model.

3.3 Mathematical Model

Under disturbed conditions, at each machine, we can write the swing equation as follows:

$$\frac{d}{dt}(W_{K.E,i}) + P_{di} = P_{mi} - P_{gi} \qquad (i=1,2,\ldots,n) \qquad (3.1)$$

where for the ith machine

$W_{K.E,i}$ = the kinetic energy of the rotor in Mega Joules

P_{di} = the damping power in MW

P_{mi} = the mechanical power input in MW

P_{gi} = the electrical power output in MW

Equation (3.1) merely states that the accelerating power for each machine is balanced by the increase in kinetic energy of the rotor and the power absorbed by the damping forces. Let us denote the synchronous frequency by f_o. The kinetic energy is proportional to the square of instantaneous frequency f_i of each machine. Hence,

$$W_{K.E,i} = W^{o}_{K.E,i}\left(\frac{f_i}{f_o}\right)^2$$

where $W^o_{K.E,i}$ is the kinetic energy of the ith machine at the synchronous frequency f_o.

$$W_{K.E,i} = W^o_{K.E,i} \frac{(f_o+\Delta f_i)^2}{f_o^2} \approx W^o_{K.E,i} \left(1 + \frac{2\Delta f_i}{f_o}\right)$$

since the deviation of f_i from f_o during a transient is small. Let α_i be the rotor angle with respect to a fixed reference and δ_i the rotor angle with respect to a synchronously rotating reference frame both in electrical radians. Thus

$$\delta_i = \alpha_i - \omega_o t$$

and

$$\frac{d\delta_i}{dt} = \frac{d\alpha_i}{dt} - \omega_o = 2\pi(f_i - f_o) = 2\pi\ \Delta f_i \stackrel{\Delta}{=} \omega_i$$

ω_i is therefore the frequency deviation from the synchronous frequency.

We shall make the assumption that the damping power P_{di} is proportional to the frequency deviation ω_i. Another form of damping is also present in the system which is due to the asynchronous torques between the machines and is proportional to the velocity difference between machines. Thus, actually we can write

$$\begin{aligned} P_{di} &= D_i\omega_i + \sum_{j=1}^{n} D_{ij}(\omega_i - \omega_j) \\ &= D_i \frac{d\delta_i}{dt} + \sum_{j=1}^{n} D_{ij} \left(\frac{d\delta_i}{dt} - \frac{d\delta_j}{dt}\right) \end{aligned} \qquad (3.2)$$

Although subsequent treatment can be carried out with the above representation of P_{di}, for ease of mathematical analysis, we will retain the D_i terms only and assume the asynchronous torques to be zero. Substituting the expression for $W_{K.E,i}$ in (3.1) and dividing by the base MVA yields

$$\frac{H_i}{\pi f_o} \frac{d^2\delta_i}{dt^2} + D_i \frac{d\delta_i}{dt} = P_{mi} - P_{gi} \qquad (i=1,2,\ldots,n) \qquad (3.3)$$

where $H_i = \frac{W^o_{K.E,i}}{\text{Base MVA}}$

is in secs, D_i is in sec/elec. radian and P_{mi} and P_{gi} are in p.u. Define $H_i/\pi f_o = M_i$. In the literature, H and M are both referred to as inertia constants. They only differ in the units, H having units of sec. and M has units of sec. per elec. radian. M is convenient in the analytic development, and hence, we retain it for our future work.

The swing equation (3.3) is therefore

$$M_i \frac{d^2\delta_i}{dt^2} + D_i \frac{d\delta_i}{dt} = P_{mi} - P_{gi} \qquad (i=1,2,...,n) \tag{3.4}$$

P_{gi} in general is a very complicated expression computed from the nonlinear differential equations of the electrical part of the synchronous machine and the algebraic equations of the transmission network and the synchronous machine [1]. Fortunately for our work, such a detailed representation of the synchronous machine is not needed. We therefore make a number of simplifying assumptions as discussed in Appendix I and summarized below which yields an analytic expression for P_{gi} in terms of the δ_i's.

1. The network is assumed to be in the sinusoidal steady state. This implies that the time constants of the transmission network are negligible compared to the electromechanical frequency of oscillation.

2. The synchronous machine is represented by a voltage source of constant magnitude determined from the steady state conditions existing prior to the fault (pre-fault load flow), in series with a reactance which is commonly called the direct axis transient reactance.

3. The phase angle of the voltage behind transient reactance coincides with the rotor angle δ_i. This important property stems from the approximate model of the synchronous machine discussed in Appendix I.

4. Loads are represented as constant impedances based on the pre-fault voltage conditions obtained from a load flow.

5. The mechanical input power P_{mi} is assumed to be constant and equal to the pre-fault value during the time interval of interest which is of the order of 1-2 sec.

With these assumptions, we now proceed to derive the relevant mathematical model.

Assume the transmission network to consist of n+m buses of which the first n buses are buses where generators are connected and at the other m buses only loads are connected. We can also have loads connected at the generator buses. The nodal admittance matrix $\underline{Y}_{BUS}$ of this network is given by

$$\underline{Y}_{BUS} = \begin{array}{c} \\ \end{array} \overset{\begin{array}{cc} n & m \end{array}}{\begin{bmatrix} \underline{Y}_1 & \underline{Y}_2 \\ \underline{Y}_3 & \underline{Y}_4 \end{bmatrix}} \begin{array}{c} n \\ m \end{array} \tag{3.5}$$

Each of the generator buses is now augmented by the generator representation consisting of a voltage source $E_i = |E_i| \angle \delta_i$ in series with the transient reactance which we shall represent by the admittance y_i (Fig. 3.1). The internal nodes of the generators are numbered 1,2,...,n and the transmission network buses n+1,...,2n+m. Prior to the fault, $|E_i|$ and δ_i are computed knowing the voltages at the generator buses and scheduled generations from the load flow data [2,3]. Deleting the voltage sources and denoting the generator admittances by the vector $\underline{y}$, we have a transmission network of 2n+m buses whose nodal admittance matrix $\hat{\underline{Y}}_{BUS}$ is given by

$$\hat{\underline{Y}}_{BUS} = \overset{\begin{array}{ccc} n & n & m \end{array}}{\begin{bmatrix} \underline{y} & -\underline{y} & \underline{0} \\ -\underline{y} & \underline{Y}_1 + \underline{y} & \underline{Y}_2 \\ \underline{0} & \underline{Y}_3 & \underline{Y}_4 \end{bmatrix}} \begin{array}{c} n \\ n \\ m \end{array} \tag{3.6}$$

We next take into consideration the loads. A constant admittance representation of the loads at the buses numbered (n+1),..,(2n+m) adds an admittance between these nodes and the ground given by

$$Y_{Li} = \frac{P_{Li} - j\,Q_{Li}}{|V_i|^2} \qquad i=n+1,...,2n+m \tag{3.7}$$

where $|V_i|$ is the magnitude of the phasor voltage at the ith bus obtained from pre-fault load flow and $P_{Li} + j\,Q_{Li}$ is the load at the ith bus. Thus, the overall modified $\underline{Y}_{BUS}$ matrix is given as (Fig. 3.2)

$$\tilde{\underline{Y}}_{BUS} = \begin{array}{c} \begin{array}{ccc} n & \qquad n & \qquad\qquad m \end{array} \\ \left[\begin{array}{ccc} \underline{y} & -\underline{y} & \underline{0} \\ -\underline{y} & \underline{Y}_1 + \underline{y} + \underline{y}_{Lg} & \underline{Y}_2 \\ \underline{0} & \underline{Y}_3 & \underline{Y}_4 + \underline{y}_{L\ell} \end{array}\right] \begin{array}{c} n \\ n \\ m \end{array} \end{array} \tag{3.8}$$

where the load admittance matrix is denoted by

$$\begin{array}{c} \begin{array}{cc} \quad n & \quad m \end{array} \\ \begin{array}{c} n \\ m \end{array} \left[\begin{array}{cc} \underline{y}_{Lg} & \underline{0} \\ \underline{0} & \underline{y}_{L\ell} \end{array}\right] \end{array} = \mathrm{Diag}(\underline{y}_{Li}) \qquad i=n+1,...,2n+m \tag{3.9}$$

So far, we have retained the topology of the network in (3.8). We now perform the matrix reduction on the $\tilde{\underline{Y}}_{BUS}$ matrix in (3.8) by eliminating all buses except the first n buses. This elimination is justified for our work since we are primarily interested in the variation of δ_i's as a function of time and not the bus voltages from the stability point of view. The $\tilde{\underline{Y}}_{BUS}$ in (3.8) is partitioned as shown below by separating the internal buses where the voltage sources are connected, from the rest of the buses where no sources are connected.

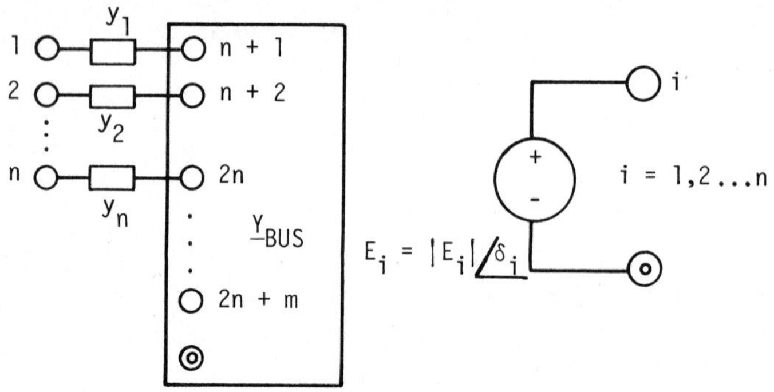

Fig. 3.1. Nodal admittance matrix $\underline{Y}_{BUS}$ of the transmission network augmented by generator representation with internal nodes 1,2, ...,n.

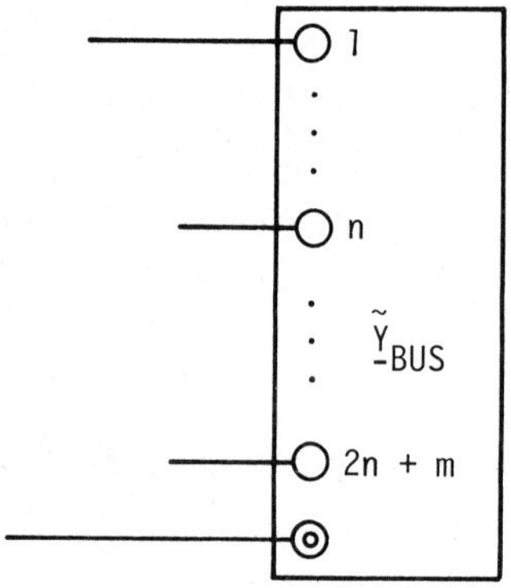

Fig. 3.2. Modified nodal admittance matrix $\underline{\tilde{Y}}_{BUS}$ for augmented network after converting loads as constant impedances (Eq. 3.8).

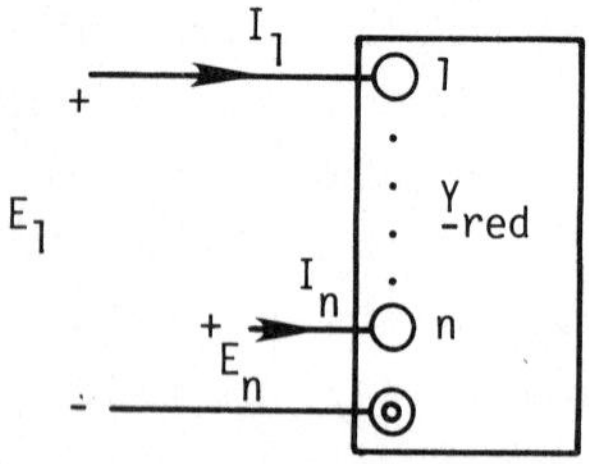

Fig. 3.3. Nodal admittance matrix $\underline{Y}_{red}$ at the internal nodes after eliminating all physical buses (Eq. 3.11).

$$\tilde{\underline{Y}}_{BUS} = \begin{array}{c} \\ \end{array}\overset{\begin{matrix} n & n+m \end{matrix}}{\begin{bmatrix} \underline{Y}_A & \underline{Y}_B \\ \underline{Y}_C & \underline{Y}_D \end{bmatrix}}\begin{matrix} n \\ n+m \end{matrix} \tag{3.10}$$

The network equations are

$$\begin{bmatrix} \underline{I}_A \\ \underline{0} \end{bmatrix} = [\tilde{\underline{Y}}_{BUS}] \begin{bmatrix} \underline{V}_A \\ \underline{V}_D \end{bmatrix}$$

Elimination of the (n+m) buses results in a $\underline{Y}_{red}$ matrix at the first n buses as $\underline{I}_A = [\underline{Y}_{red}]\underline{V}_A$ where

$$\underline{Y}_{red} = [\underline{Y}_A - \underline{Y}_B\, \underline{Y}_D^{-1}\, \underline{Y}_C] \tag{3.11}$$

This $\underline{Y}_{red}$ matrix is the internal bus description of the system (Fig. 3.3) and although it masks the topological aspects, it simplifies the analysis considerably in terms of obtaining an analytic expression for P_{gi} in Eq. (3.3) in terms of δ_i's. Let elements of this $\underline{Y}_{red}$ matrix be denoted by Y_{ij}. Now we have the multiport description at the internal nodes as shown in Fig. (3.3) with voltage sources connected across the ports. The expression for real power generation is

$$P_{gi} = \mathrm{Re}[E_i\, I_i^*] = \mathrm{Re}[E_i \sum_{j=1}^{n} Y_{ij}^*\, E_j^*] \quad i=1,2,\ldots,n.$$

Let

$$E_i = |E_i|\underline{/\delta_i},\ Y_{ij} = G_{ij} + j\, B_{ij} = |Y_{ij}|\,\underline{/\phi_{ij}}$$

where $G_{ij} = |Y_{ij}|\cos\phi_{ij}$ and $B_{ij} = |Y_{ij}|\sin\phi_{ij}$

Then

$$P_{gi} = |E_i|^2 G_{ii} + \sum_{\substack{j=1 \\ \neq i}}^{n} |E_i||E_j||Y_{ij}[\cos(\phi_{ij}-(\delta_i-\delta_j))]$$

$$= |E_i|^2 G_{ii} + \sum_{\substack{j=1 \\ \neq i}}^{n} |E_i||E_j||Y_{ij}|$$

$$[\cos\phi_{ij}\cos(\delta_i-\delta_j) + \sin\phi_{ij}\sin(\delta_i-\delta_j)] \qquad (3.12)$$

Define

$$|E_i||E_j||Y_{ij}|\cos\phi_{ij} = |E_i||E_j|G_{ij} \triangleq D_{ij}$$

$$|E_i||E_j||Y_{ij}|\sin\phi_{ij} = |E_i||E_j|B_{ij} \triangleq C_{ij}$$

Then

$$P_{gi} = |E_i|^2 G_{ii} + \sum_{\substack{j=1 \\ \neq i}}^{n} [C_{ij}\sin(\delta_i-\delta_j) + D_{ij}\cos(\delta_i-\delta_j)]$$

Substituting (3.12) in (3.4), we get

$$M_i \frac{d^2\delta_i}{dt^2} + D_i \frac{d\delta_i}{dt} = P_i - P_{ei} \qquad i=1,2,\ldots,n \qquad (3.13)$$

where $P_i = P_{mi} - |E_i|^2 G_{ii}$

$$P_{ei} = \sum_{\substack{j=1 \\ \neq i}}^{n} [C_{ij}\sin(\delta_i-\delta_j) + D_{ij}\cos(\delta_i-\delta_j)]$$

Equation (3.13) is the basic form of the so-called swing equation for each machine. Prior to the disturbance, the power system is in a steady state with δ_i assuming constant values δ_i^o and $\dot{\delta}_i = 0$ for all i. At the instant of fault, $t = 0^+$, these initial conditions are valid since neither the rotor angles nor the frequency changes instantly. We have two sets of differential equations for $t > 0$, one for the faulted state until $t = t_{cl}$ and the other for $t > t_{cl}$. The difference between these two sets of differential equations is in the transmission configuration, this being reflected in the Y_{ij}

parameters of the $\underline{Y}_{red}$ matrix, i.e., it effects P_{gi} only. We can symbolically therefore write

Faulted State

$$M_i \frac{d^2\delta_i}{dt^2} + D_i \frac{d\delta_i}{dt} = P_{mi} - P_{gi}^f \qquad 0 < t \leq t_{cl} \qquad (3.14)$$

$$= P_i^f - P_{ei}^f \qquad \delta_i(0) = \delta_i^o,\ \dot{\delta}_i(0) = 0$$

$$i=1,2,\ldots,n.$$

Post Fault State

$$M_i \frac{d^2\delta_i}{dt^2} + D_i \frac{d\delta_i}{dt} = P_{mi} - P_{gi}^{pf} \qquad t \geq t_{cl},\ \delta_i(t_{cl}),$$

$$= P_i^{pf} - P_{ei}^{pf} \qquad \dot{\delta}_i(t_{cl}) \qquad (3.15)$$

determined from (3.14)

$i=1,2,\ldots,n$

The problem of stability can be stated as follows. "Given t_{cl}, does $\delta_i(t)$ assume constant values and $\dot{\delta}_i(t) \to 0$ as $t \to \infty$? If it does, the system is then stable. Otherwise, the system is unstable." The problem of finding t_{cr} is to find the maximum value of t_{cl} so that the system is stable.

We at once notice the mathematical ramifications of the above stated problem. We have basically two sets of d.e's (3.14) and (3.15) which have a discontinuity at $t = t_{cl}$ and we are asking not for stability of the origin or the equilibrium point of (3.14) but rather boundedness of the trajectory defined by (3.14) and (3.15). This is a rather difficult problem to solve except by simulation. To find t_{cr}, simulation has to be repeated for different values of t_{cl} so that we finally find a t_{cl} such that the system is stable for $t_{cl}-\varepsilon$ and unstable for $t_{cl}+\varepsilon$ where $\varepsilon > 0$ and small. Such a value of t_{cl} is called the critical clearing time t_{cr}. Repeated simulation to find t_{cr} for a single fault can be expensive computationally since in a large system there are several faults to be considered. Typically, an experienced engineer does 3-4 repetitive simulations for a single fault before he can find t_{cr}.

Lyapunov's method addresses itself to the question of avoiding this repetitive simulation and having just one simulation to find t_{cr} directly. If t_{cr} were to be estimated thus accurately, it will mean a substantial savings in computer time for planning studies.

Since initial conditions for the post fault system (3.15) (at $t = t_{cl}$) correspond to the terminal values of δ_i and $\dot{\delta}_i$ from the solution of the faulted system (3.14) at $t = t_{cl}$, it would seem logical to construct a region of attraction around the post fault equilibrium point and for stability, examine if the terminal state of the faulted system (3.14) lies within the region of attraction of the post-fault system (3.15). Theorem 2.7.5 provides a basis for estimating the region defined by the inequality $V(\underline{x}) < C$ for the post-fault system such that inside the region $V(\underline{x}) > 0$ and $\dot{V}(x) < 0$ or $\dot{V}(\underline{x}) \leq 0$ providing the points where $\dot{V}(\underline{x}) = 0$ are not solutions of the post fault system. The faulted equations are then integrated until $V(\underline{x}) = C$ and the instant of time when this equality is satisfied is an estimate of t_{cr}. Since Lyapunov's method yields sufficient conditions, the region defined by $V(\underline{x}) < C$ is contained in the exact region of stability and hence the computation will yield in general values of t_{cr} less than the actual value. There are two issues to be resolved. The first one is how to construct a region of stability for a given fault (disturbance) and the manner in which it is cleared, i.e. with or without line switching. The second one is regarding the construction of the Lyapunov function $V(\underline{x})$ itself since different Lyapunov functions will yield different values of C. Computation of C, so that accurate estimates of t_{cr} may be obtained, has been a major bottleneck in the practical implementation of Lyapunov's method. Initial research tended to compute a single value of C from (3.15) only for all fault locations and ignored the faulted equation (3.14) totally. Since the mode of instability is determined by the fault and the faulted dynamics (3.14), it is natural that a computation of C which ignored (3.14) will yield conservative values of t_{cr} for certain faults and acceptable values for other faults. This puzzled research workers for a long time until very recently when the work reported in Refs. [23-25] of Chapter I removed this conservativeness.

3.4 State Space Models

We now develop the appropriate state space models for (3.13). The parameters will pertain either to the faulted state or the post-fault state depending respectively on whether the equations are to be integrated or a Lyapunov function has to be devloped.

In the literature, there are several types of models proposed and often there is difficulty in reconciling results obtained from different models. One school of thought prefers to treat Eq. (3.13) as it is in which case the post-fault stable equilibrium point (SEP) is at a point other than the origin in the state space and the Lyapunov function is not zero at the origin. The other type of model transfers the post-fault SEP to the origin by a coordinate transformation and $V(\underline{x})$ is zero at the origin. Moreover, the equations in this case can be put in the standard Luré form well known in modern stability theory and control theoretic approaches are possible. We follow the latter approach. In all power system problems, a reference angle is always required. We can have the state space model based on machine angle as a reference (MAR) or center of angle as a reference (COA). We discuss both types of models in this chapter and point out the connection between them.

3.4.1 State Space Model Using Machine Angle as Reference

Define

$$\underline{x}^* = [\omega_1\omega_2,\ldots,\omega_n \,\vdots\, \delta_1\delta_2,\ldots,\delta_n]^T$$

Then, Eq. (3.13) is equivalent to the following 2n equations.

$$\dot{\omega}_i = -\frac{D_i}{M_i}\omega_i + P_i - P_{ei}$$

$$\dot{\delta}_i = \omega_i \qquad i=1,2,\ldots,n \tag{3.16}$$

This mode has 2n state variables with rotor angles δ_i and rotor speed deviations ω_i as state variables. This was the model used in earlier works in Refs. [13] and [15] of Chapter I in deriving the Lyapunov functions for multi-machine power systems using Moore-Anderson Theorem and Kalman's construction procedure respectively. While these Lyapunov functions are valid if we invoke the partial stability concept [4], they are not valid in the state space defined by $\underline{x}^*$ if stability is considered in the sense of Lyapunov. This was pointed out by Sastry and Murthy [5] and Ribbens-Pavella (Ref. [19], Chap. I). P_{ei}'s are functions of angular differences only and, hence, we must take angular differences of machines with respect to a reference machine as the angle subvector while the velocity subvector remains unchanged. (Refs. [10,16], Chap. I) Thus, the proper description of (3.13) is through the (2n-1) dimensional state vector

$$\underline{x} = [\omega_1,\omega_2,\ldots,\omega_n \,|\, \delta_1-\delta_n,\delta_2-\delta_n,\ldots,\delta_{n-1}-\delta_n]^T$$

where the nth machine has been taken as the reference machine. This results in the state space model

$$\begin{aligned}\dot{\omega}_i &= -\frac{D_i}{M_i}\omega_i + P_i - P_{ei} && i=1,2,\ldots,n\\ \dot{\delta}_i - \dot{\delta}_n &= \omega_i - \omega_n && i=1,2,\ldots,n-1\end{aligned} \tag{3.17}$$

Now let us consider the case of uniform damping ($\frac{D_i}{M_i} = \lambda$ for all i) or zero damping case ($D_i = 0$ for all i). The state variables will now be

$$\underline{x} = [\omega_1-\omega_n,\omega_2-\omega_n,\ldots,\omega_{n-1}-\omega_n \,|\, \delta_1-\delta_n,\delta_2-\delta_n\ldots\delta_{n-1}-\delta_n]^T$$

The state equations will be

$$\frac{d}{dt}(\omega_i-\omega_n) = -\lambda(\omega_i-\omega_n) + \left(\frac{P_i-P_{ei}}{M_i}\right) - \left(\frac{P_n-P_{en}}{M_n}\right)$$

$$i=1,2,\ldots,n-1 \tag{3.18}$$

$$\frac{d}{dt}(\delta_i-\delta_n) = (\omega_i-\omega_n) \qquad i=1,2,\ldots,n-1$$

For the zero damping case $\lambda \equiv 0$. Thus, both for uniform and zero damping case, the order of the state space model is 2n-2. In the literature, there has been considerable discussion regarding the correct order of the state space model for cases of non-uniform damping, uniform damping and zero damping [6]. While in the preceding discussion, we have resorted to purely physical arguments, the derivation of the correct order of the state model has also been done via i) controllability and observability notion [5], and ii) minimal realization theory [7]. This will be explained in the context of Luré formulation in section 3.6.

3.5 State Space Model in Center of Angle Reference Frame

An alternate way of viewing the model is through what is called the center of inertia or center of angle (COA) formulation. This was proposed by Tavora and Smith [8], Stanton [9] and Fouad and Lugtu [10]. As we shall see in Chapter VI, this formulation is particularly suited for the energy function approach. The formulation is analogous to the center of mass principle in mechanics. Instead of considering δ_i with respect to a synchronously rotating reference frame, we have a reference frame which is time varying.

Consider the swing equations (3.13) and adopt the compact notation $\ddot{\delta} = \frac{d^2\delta}{dt^2}$ and $\dot{\delta} = \frac{d\delta}{dt}$. If we add all the n equations in (3.13), we get

$$\sum_{i=1}^{n} M_i \ddot{\delta}_i + \sum_{i=1}^{n} D_i \dot{\delta}_i = \sum_{i=1}^{n} P_i - \sum_{i=1}^{n} P_{ei} \tag{3.19}$$

$$i=1,2,\ldots,n$$

Define the center of angle δ_o as

$$\delta_o = \frac{1}{M_T} \sum_{i=1}^{n} M_i \delta_i \tag{3.20}$$

where

$$M_T = \sum_{i=1}^{n} M_i$$

Also define the new rotor angle θ_i with respect to δ_o as $\theta_i = \delta_i - \delta_o$. Let $\dot{\theta}_i = \tilde{\omega}_i = \omega_i - \omega_o$ where $\omega_o = \dot{\delta}_o$. Note that not all θ_i's and $\tilde{\omega}_i$'s are linearly independent since from the definition of θ_i it follows that

$$\sum_{i=1}^{n} M_i \theta_i = 0 \quad \text{and} \quad \sum_{i=1}^{n} M_i \dot{\theta}_i = 0 \tag{3.21}$$

Equations in the new coordinates are obtained after some manipulation as

$$M_T \ddot{\delta}_o = -D_T \dot{\delta}_o - \sum_{i=1}^{n} D_i \dot{\theta}_i + \sum_{i=1}^{n} P_i - 2 \sum_{i=1}^{n} \sum_{j=i+1}^{n} D_{ij} \cos\delta_{ij} \tag{3.22}$$

where $D_T = \sum_{i=1}^{n} D_i$

Define $P_{COA} = \sum_{i=1}^{n} P_i - 2 \sum_{i=1}^{n} \sum_{j=i+1}^{n} D_{ij} \cos\delta_{ij}$ (3.23)

Then $M_T \ddot{\delta}_o = -D_T \dot{\delta}_o - \sum_{i=1}^{n} D_i \dot{\theta}_i + P_{COA}$ (3.24)

The dynamics of the center of angle are governed by (3.24). The system equations in the variables θ_i now become [11] (after algebraic manipulation)

$$M_i \ddot{\theta}_i = -(D_i \dot{\theta}_i - \frac{M_i}{M_T} \sum_{i=1}^{n} D_i \dot{\theta}_i) - (D_i - \frac{M_i}{M_T} D_T) \dot{\delta}_o + P_i - P_{ei} - \frac{M_i}{M_T} P_{COA} \qquad i=1,2,\ldots,n \tag{3.25}$$

Equations (3.24) and (3.25) therefore represent a set of coupled (n+1) second order differential equations.

At the end of the transient period, if the system settles to a new steady state, we should have $\ddot{\delta}_o = 0$, $\ddot{\theta}_i = 0$, $\dot{\theta}_i = 0$. The new system angular frequency is $\omega = \omega_o + \dot{\delta}_{os}$ where $\dot{\delta}_{os}$ is the equilibrium value of $\dot{\delta}_o$ obtained from (3.24) as

$$D_T\dot{\delta}_o = P_{COA} \tag{3.26}$$

and ω_o is the inertial angular frequency. As we shall see in Sec. 3.6, at the post-fault SEP, $P_{COA} = 0$ and, hence, $\dot{\delta}_{os} = 0$ so that the system regains the original frequency.

3.5.1 Simplifications of COA Equations for Uniform and Zero Damping Case

The COA formulation gets simplified in the case of either uniform damping or zero damping.

For the uniform damping case, we have the equation (3.24) for center of angle as

$$M_T\ddot{\delta}_o = -D_T\dot{\delta}_o + P_{COA} \tag{3.27}$$

since $\sum_{i=1}^{n} D_i\dot{\theta}_i = 0$ by virtue of Eq. (3.21).

For zero damping case, (3.27) becomes

$$M_T\ddot{\delta}_o = P_{COA} \tag{3.28}$$

The machine equations (3.25) for the uniform damping case simplify to

$$M_i\ddot{\theta}_i = -D_i\dot{\theta}_i + P_i - P_{ei} - \frac{M_i}{M_T} P_{COA} \qquad i=1,2,\ldots,n \tag{3.29}$$

For the zero damping case, we set $D_i \equiv 0$ for all i.

The state space model in 2n dimension space with the state variables as $[\omega_1,\ldots,\omega_n \mid \theta_1,\ldots,\theta_n]$ for the uniform damping case is obtained from (3.29) as

$$M_i \dot{\tilde{\omega}}_i = -D_i \tilde{\omega}_i + P_i - P_{ei} - \frac{M_i}{M_T} P_{COA}$$
$$\dot{\theta}_i = \tilde{\omega}_i \qquad (3.30)$$

For the zero damping case, $D_i \equiv 0$ for all i.

Equation (3.30) is decoupled from (3.27) or (3.28) and for both uniform and zero damping cases the correct dimension of the state space is (2n-2). Analogous to the machine angle reference case, we can take the nth machine as reference machine and write the state space equations from (3.30) as

$$\dot{\tilde{\omega}}_i - \dot{\tilde{\omega}}_n = -\lambda(\tilde{\omega}_i - \tilde{\omega}_n) + \left(\frac{P_i}{M_i} - \frac{P_n}{M_n}\right) - \left(\frac{P_{ei}}{M_i} - \frac{P_n}{M_n}\right)$$
$$\dot{\theta}_i - \dot{\theta}_n = \tilde{\omega}_i - \tilde{\omega}_n \qquad i=1,2,\ldots,n-1 \qquad (3.31)$$

For zero damping case, $\lambda \equiv 0$.

It is not surprising that these equations are identical to (3.18) since $\tilde{\omega}_i - \tilde{\omega}_n = \omega_i - \omega_n$ and $\theta_i - \theta_n = \delta_i - \delta_n$. Thus, the state space representation for the uniform and zero damping case is the same whether we take the machine angle or center of angle as reference angle. In the case of non-uniform damping, however, the COA reference equations (3.25) do not lend themselves to a neat representation.

3.6 State Space Models with Post-Fault SEP Transferred to Origin (Machine Angle Reference)

The preceding discussion has focussed on different formulations of the state space equations used by many research workers in the literature who have constructed Lyapunov functions based on energy considerations. For Lyapunov stability work, we now cast the equations in a form so that the origin is the equilibrium point. This form is necessary if we wish to construct Lyapunov functions in a systematic manner by any of the methods discussed in Chapter II.

As discussed in Section 3.3, we seek to construct a region of stability around the post-fault stable equilibrium point.

Hence, the differential equations that we must consider are those corresponding to the post-fault state given by (3.15). The $\underline{Y}_{red}$ matrix in (3.11) must be calculated for this case to obtain the Y_{ij}'s. If there is no line switching following a fault, the post-fault SEP is the same as the pre-fault operating point and the $\underline{Y}_{red}$ matrix is the same as for the pre-fault configuration. Let us consider the general case when the two configurations are different due to line switching.

The equilibrium solution of the post-fault system (3.15) is obtained from the equations

$$\frac{d\delta_i}{dt} = \omega_i = 0 \qquad (3.32)$$

$$i=1,2,\ldots,n$$

$$P_i^{pf} = P_{ei}^{pf} \qquad (3.33)$$

Equation (3.33) are n equations in (n-1) unknowns, namely the angle differences. Hence, we may solve only the first (n-1) equations for the angle differences while the remaining equation can be interpreted as a constraining equation. In general, we can write the constraint by summing up all the n equations in (3.33), i.e. $\sum_{i=1}^{n} (P_i^{pf} - P_{ei}^{pf}) = 0$. Since the sine terms in this expression cancel out, the constraint becomes $P_{COA} = 0$ where P_{COA} is defined by (3.23) with the Y_{ij} parameters pertaining to the post-fault configuration. In terms of COA notation, this implies $\dot{\delta}_{os} = 0$, in (3.26), i.e., the system regains the original frequency. In the case of zero transfer conductances, the condition $P_{COA} = 0$ becomes $\sum_{i=1}^{n} P_i = 0$, i.e., $\sum_{i=1}^{n} (P_{mi} - |E_i|^2 G_{ii}) = 0$. This condition for existence of a solution of (3.33) if very often quoted in the literature. Let the solution for Eq. (3.33) be δ_i^s. In solving (3.33), one of the angles δ_i generally the rotor angle corresponding to slack bus, is taken as reference and equal to its pre-fault value. We are seeking a stable operating solution and, hence δ_i^s will be in the neighborhood of the pre-fault equilibrium point δ_i^o. The post-fault SEP is therefore $\omega_i = 0$, $\delta_i = \delta_i^s$ $(i=1,2,\ldots,n)$.

Since the solution to the nonlinear equations (3.33) is δ_i^s, we have using (3.12),

$$P_{mi} - |E_i|^2 G_{ii} = \sum_{\substack{j=1\\ \neq i}}^{n} |E_i||E_j||Y_{ij}|\cos(\phi_{ij} - (\delta_i^s-\delta_j^s))$$

$$= \sum_{\substack{j=1\\ \neq i}}^{n} [C_{ij}\sin(\delta_i^s-\delta_j^s) + D_{ij}\cos(\delta_i^s-\delta_j^s)] \tag{3.34}$$

$$= P_{ei}(\underline{\delta}^s)$$

Hence, the swing equations (3.13) can be written for the post-fault state as

$$M_i \frac{d^2\delta_i}{dt^2} + D_i \frac{d\delta_i}{dt^2} = P_{ei}(\underline{\delta}^s) - P_{ei}(\underline{\delta})$$

$$= \sum_{\substack{j=1\\ \neq i}}^{n} E_i E_j Y_{ij}\{\cos[\phi_{ij} - (\delta_i^s-\delta_j^s)] \tag{3.35}$$

$$- \cos[\phi_{ij} - (\delta_i-\delta_j)]\} \qquad i=1,2,\ldots,n$$

For ease of notation, the magnitude sign $|\cdot|$ has been dropped from E_i, E_j and Y_{ij}. Define the state vector as

$$\underline{x}^* = [\omega_1,\omega_2,\ldots,\omega_n \mid \delta_1-\delta_1^s,\delta_2-\delta_2^s,\ldots,\delta_n-\delta_n^s]^T$$

With these variables (3.35) can be reduced to the form

$$\underline{x}^* = \underline{A}\,\underline{x}^* - \underline{B}^*\,\underline{f}(\underline{C}^*\underline{x}^*) \tag{3.36}$$

This model is the equivalent of (3.16) with origin as the equilibrium point. However, we must ensure that (3.36) is in the minimal state space.

3.6.1 Minimal Order of State Space Model

The question has been investigated in the literature from different angles viz.

1. Physical argumentation: This was elaborated in Sec. 3.4.
2. Controllability and observability notions [8].
3. Minimal realization concept [9].

However, all of these ways of looking at the problem lead to the same ultimate result, namely

i) for the non-uniform damping case, i.e.

$$\frac{D_1}{M_1} \neq \frac{D_2}{M_2} \neq \frac{D_3}{M_3} \;\ldots\; \neq \frac{D_n}{M_n}$$

the minimal dimension of the state space is 2n-1.

ii) for the uniform damping case, i.e.

$$\frac{D_1}{M_1} = \frac{D_2}{M_2} = \frac{D_3}{M_3} = \ldots = \frac{D_n}{M_n} = \lambda$$

and zero damping case, i.e. $D_i \equiv 0$ for all i, the minimal dimension of the state space is 2n-2.

The discussion is facilitated by assuming a model with zero transfer conductances. With this assumption, $\phi_{ij} = \pi/2$, i.e. $D_{ij} = 0$ in (3.34) and (3.35). Equations (3.35) now become

$$M_i \frac{d^2\delta_i}{dt^2} + D_i \frac{d\delta_i}{dt} = \sum_{\substack{j=1 \\ \neq i}}^{n} E_i E_j B_{ij} [\sin(\delta_i^s - \delta_j^s) - \sin(\delta_i - \delta_j)] \qquad (3.37)$$

The state variables are defined by

$$\underline{x}^* = [\omega_1, \omega_2, \ldots \omega_n \,\vdots\, \delta_1 - \delta_1^s, \ldots, \delta_n - \delta_n^s]^T$$

The state model becomes

$$\dot{\underline{x}}^* = \underline{A}\,\underline{x}^* - \underline{B}\,\underline{f}(\underline{\sigma}) \qquad (3.38)$$

$$\underline{\sigma} = \underline{C}\,\underline{x}^*$$

$$
\underline{A} = \begin{array}{c} \begin{matrix} n & \qquad n \end{matrix} \\ \begin{bmatrix} -\underline{D}\,\underline{M}^{-1} & \underline{0} \\ \underline{I} & \underline{0} \end{bmatrix} \end{array} \begin{matrix} n \\ n \end{matrix} \qquad\qquad \underline{B} = \begin{array}{c} n \\ \begin{bmatrix} \underline{M}^{-1}\underline{K} \\ \underline{0} \end{bmatrix} \end{array} \begin{matrix} n \\ m \end{matrix}
$$

$$
\underline{C} = \begin{array}{c} \begin{matrix} n & \quad n \end{matrix} \\ [\,\underline{0} \quad \underline{K}^T] \end{array} \qquad\qquad m = \frac{n(n-1)}{2}
$$

$$
\underline{D} = \mathrm{Diag}(D_i); \quad \underline{M} = \mathrm{Diag}(M_i)
$$

$$
\underline{K} = [\underline{K}_1 \,\vdots\, \underline{K}_2 \;\cdots\; \vdots\, \underline{K}_{n-1}]
$$

$$
\begin{array}{c}
\begin{matrix} \qquad n-1 \qquad\qquad & \qquad n-2 \qquad\qquad & \qquad n-3 \qquad\qquad & \end{matrix} \\
= \left[\begin{array}{ccc|ccc|cc|c|c}
1 & 0 \dots & 0 & 1 & 1 \dots & 1 & 0 \dots & 0 & \dots & 0 \\
0 & 1 & . & -1 & 0 & 0 & 1 & 1 & & 0 \\
. & . & . & . & . & . & -1 & . & & . \\
. & . & . & . & . & . & & . & & . \\
. & . & . & . & . & . & 0 & . & \dots & \\
. & . & . & . & . & . & . & . & & 1 \\
. & . & . & . & . & . & . & . & & . \\
. & . & 1 & . & . & -1 & . & -1 & & -1 \\
-1 & -1 & -1 & 0 & 0 & 0 & 0 & 0 & & 0
\end{array}\right] n
\end{array}
$$

$$
= \left[\begin{array}{c|c} \underline{K}_1 & \begin{array}{c} \underline{J} \\ \hline \underline{0} \end{array} \end{array}\right]
$$

$\underline{f}(\underline{\sigma})$ = a vector of dimension m $(= \frac{n(n-1)}{2})$ whose i^{th} component is

$$
f_i(\sigma_i) = E_p E_q B_{pq}[\sin(\sigma_i + \beta_i) - \sin\beta_i]
$$

with $\quad \underline{\beta} = \underline{K}^T \underline{\delta}^s$

The system (3.38) can be put in the block diagram as in Fig. 3.4. The transfer function of the linear part is given by

$$\frac{\underline{\sigma}(s)}{-\mathcal{L}[\underline{f}(\underline{\sigma})]} \triangleq \underline{W}_N(s) = \underline{C}(s\,\underline{I} - \underline{A})^{-1}\,\underline{B}$$

$$= \frac{1}{s}\,[\underline{K}^T(s\,\underline{I} + \underline{\lambda})^{-1}\,\underline{M}^{-1}\,\underline{K}] \qquad (3.39)$$

where $\underline{\lambda} = \mathrm{Diag}(\lambda_1, \lambda_2, \ldots \lambda_n) = \underline{D}\underline{M}^{-1}$

The three cases of interest are

i) non-uniform damping ($\lambda_i \neq \lambda_j$ for $i \neq j$)
ii) uniform damping $\lambda_i = \lambda$ for all i
iii) zero damping $\lambda_i = 0$ for all i

In the case of uniform damping, the transfer function matrix (3.39) becomes

$$\underline{W}_u(s) = \frac{1}{s(s+\lambda)}\,\underline{K}^T\underline{M}^{-1}\underline{K}$$

The zero damping is a special case of $\underline{W}_u(s)$ with $\lambda \equiv 0$ and the transfer function is $\underline{W}_o(s) = \frac{1}{s^2}\,\underline{K}^T\underline{M}^{-1}\underline{K}$.

3.6.2 Minimal Realization

Physically speaking, we would like to ensure that the number of state variables in (3.38) is a minimal set giving rise to the transfer function matrix $\underline{W}_N(s)$. If indeed it is the case, then the corresponding ($\underline{A}$, $\underline{B}$, $\underline{C}$) is called the minimal realization of $\underline{W}_N(s)$. We, therefore, test for the minimal order of the state space model. The minimal order of the state space model is given by the degree of the rational transfer function $\underline{W}_N(s)$, $\underline{W}_u(s)$ or $\underline{W}_o(s)$ as the case may be. It has been shown by Gilbert [12] that if $\underline{W}(s)$ has a partial fraction expansion of the form

$$\underline{W}(s) = \sum_{i=1}^{n} \underline{Z}_i(s+\lambda_i)^{-1}$$

then the degree of $\underline{W}(s)$ is equal to the sum of the ranks of $\underline{Z}_i$. If $\underline{W}(s) = a(s)\underline{C}$ with a(s) a rational scalar and $\underline{C}$ a constant matrix, then the degree of $\underline{W}(s)$ is equal to degree of a(s) times the rank of $\underline{C}$ where degree of a(s) is the highest power of s in the numerator or denominator of a(s).

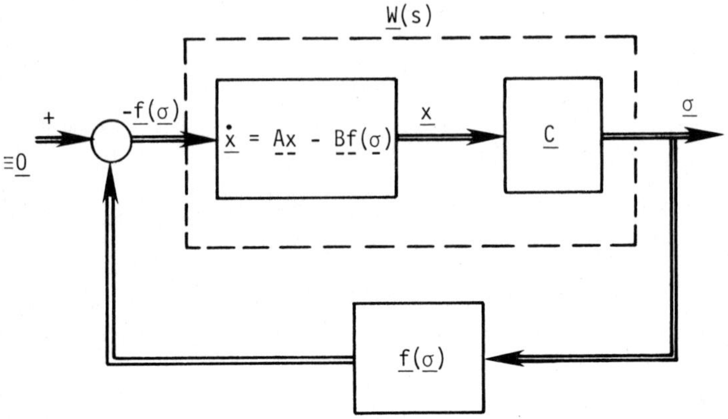

Fig. 3.4. Block diagram representation of a multi-machine power system model.

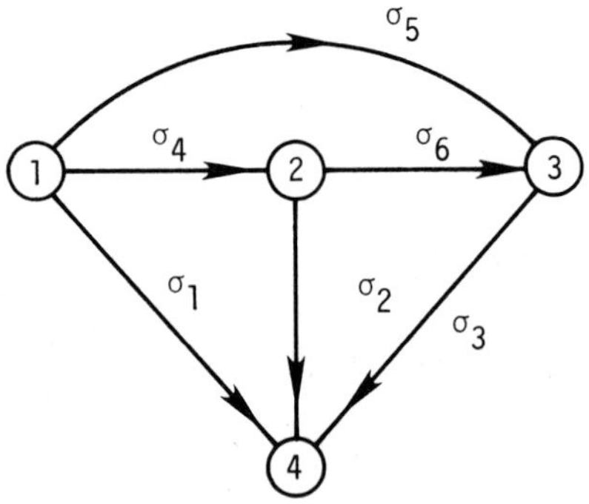

Fig. 3.5. Oriented graph for the internal node description of angle variables.

Non-uniform Damping

$\underline{W}_N(s)$ can be rewritten as

$$\underline{W}_N(s) = \frac{1}{s} \underline{K}^T \underline{D}^{-1} \underline{K} + \sum_{i=1}^{n} \frac{1}{(s+\lambda_i)} \underline{K}^T \underline{F}_i \underline{K}$$

where $\underline{F}_i$ is an n×n matrix having all zero elements except the ii element which is $-\frac{1}{D_i}$. The degree of $\underline{W}_N(s)$ = Rank of $(\underline{K}^T \underline{D}^{-1} \underline{K}) + \sum_{i=1}^{n}$ Rank of $(\underline{K}^T \underline{F}_i \underline{K})$. The rank of $\underline{K}^T \underline{F}_i^{-1} \underline{K}$ is unity for each i and that of $\underline{K}^T \underline{D}^{-1} \underline{K}$ is n-1. Hence, degree of $\underline{W}_N(s)$ = 2n-1 which is therefore the minimal order of the state space for the system with non-uniform damping.

Let the triple $[\underline{A}, \underline{B}, \underline{C}]$ constitute the minimal realization for the n machine power system with non-uniform damping. The following state variables spanning the minimal state space are chosen

$$x_i = \omega_i = \frac{d\delta_i}{dt} \qquad\qquad i=1,2,\ldots,n$$

$$x_{n+i} = (\delta_i - \delta_n) - (\delta_i^s - \delta_n^s) \qquad\qquad i=1,2,\ldots,n$$

The state model is given by

$$\dot{\underline{x}} = \underline{A}\,\underline{x} - \underline{B}\,\underline{f}(\underline{\sigma})$$
$$\underline{\sigma} = \underline{C}\,\underline{x}$$

$$(3.40)$$

$$\underline{A} = \begin{array}{c} \\ \end{array} \begin{array}{cc} n & n-1 \\ \left[\begin{array}{cc} -\underline{D}\,\underline{M}^{-1} & \underline{0} \\ \underline{K}_1^T & \underline{0} \end{array}\right] \end{array} \qquad \underline{B} = \begin{array}{cc} \quad m \\ \left[\begin{array}{c} \underline{M}^{-1}\underline{K} \\ \underline{0} \end{array}\right] \begin{array}{c} n \\ n-1 \end{array} \end{array}$$

$$\underline{C} = \begin{array}{c} n \quad n-1 \\ [\, \underline{0} \quad \underline{S}\,]m \end{array} \quad = \begin{array}{c} n \qquad n-1 \\ \left[\begin{array}{cc} \underline{0} & \underline{I} \\ \underline{0} & \underline{J}^T \end{array}\right] \begin{array}{c} n-1 \\ (m-n+1) \end{array} \end{array}$$

$$\underline{K}_1^T \underline{S} = \underline{K}$$

It may be easily verified that the transfer function $\underline{W}_N(s)$ computed with the above state space description is identical to (3.39).

The preceding discussion can be readily extended to the uniform and zero damping cases. We will only derive the minimal order of the state space. The minimal realizations are discussed in Ref. [7].

Uniform Damping

The transfer function matrix $\underline{W}_u(s)$ is

$$\underline{W}_u(s) = \frac{1}{s(s+\lambda)} \underline{K}^T\underline{M}^{-1}\underline{K}$$

$$= \frac{1}{s\lambda} \underline{K}^T\underline{M}^{-1}\underline{K} - \frac{1}{(s+\lambda)\lambda} \underline{K}^T\underline{M}^{-1}\underline{K}$$

Since the rank of $\underline{K}^T\underline{M}^{-1}\underline{K}$ is (n-1), the degree of $\underline{W}_u(s)$ is 2n-2 and, hence, it is the minimal order of the state space.

Zero Damping

The transfer function matrix is

$$\begin{aligned} \underline{W}_o(s) &= \frac{1}{s^2} \underline{K}^T\underline{M}^{-1}\underline{K} \\ &= a(s)\, \underline{K}^T\underline{M}^{-1}\underline{K} \end{aligned} \tag{3.42}$$

Since the degree of a(s) = 2 and the rank of $\underline{K}^T\underline{M}^{-1}\underline{K}$ = n-1, the degree of $\underline{W}_o(s)$ = 2(n-1) = 2n-2 which is the minimal order of of the state space.

3.6.3 State Space Model with Non-negligible Transfer Conductances (Non-uniform Damping)

We will now illustrate the state space equations including transfer conductances for a 4-machine case and then generalize for an n machine case with non-uniform damping. The reason for choosing a 4-machine system is that the structure of various matrices becomes clearer than in a 3-machine case.

Four Machine System

The state variables are

$$x = [\omega_1,\omega_2,\omega_3,\omega_4 \mid \delta_{14}-\delta_{14}^s,\delta_{24}-\delta_{24}^s,\delta_{34}-\delta_{34}^s]^T$$

where $\delta_{i4} = \delta_i - \delta_4 \qquad (i=1,2,3)$

The formulation is facilitated by the introduction of auxiliary variables σ_i's as follows.

$$\sigma_1 = \delta_{14} - \delta_{14}^s, \quad \sigma_2 = \delta_{24} - \delta_{24}^s, \quad \sigma_3 = \delta_{34} - \delta_{34}^s$$

$$\sigma_4 = \delta_{12} - \delta_{12}^s, \quad \sigma_5 = \delta_{13} - \delta_{13}^s, \quad \sigma_6 = \delta_{23} - \delta_{23}^s$$

Note the manner in which the σ_i's are ordered. The first (n-1) σ_i's are the state variables x_i (i=5,6,7) themselves corresponding to the relative rotor angles. The others are ordered so that for i > n-1

$$\sigma_i = \delta_{pq} - \delta_{pq}^s \qquad p=1,2,\ldots \qquad q=2,3,\ldots \qquad (q > p)$$

The rule for numbering can also be deduced from Fig. 3.5 where every internal bus is connected to every other bus and bus 4 is the reference bus.

Expanding (3.35) and rearranging the terms, the state model can be derived as:

$$\underline{\dot{x}} = \underline{A}\underline{x} - \underline{B}_1\underline{f}(\underline{\sigma}) - \underline{B}_2\underline{g}(\underline{\sigma})$$

$$\underline{\sigma} = \underline{C}\underline{x} \tag{3.43}$$

$$\underline{x} = (x_1x_2x_3x_4 \mid x_5x_6x_7)^T, \qquad \underline{\sigma} = (\sigma_1\sigma_2\sigma_3 \mid \sigma_4\sigma_5\sigma_6)^T$$

Let $\lambda_i = D_i/M_i$. The matrices $\underline{A}$, $\underline{B}_1$, $\underline{B}_2$, $\underline{C}$ and the nonlinearities $f_i(\underline{\sigma}) = f_i(\sigma_i)$ and $g_i(\underline{\sigma}) = g_i(\sigma_i)$ are shown below:

$$\underline{A} = \left[\begin{array}{cccc|c} -\lambda_1 & 0 & 0 & 0 & \\ 0 & -\lambda_2 & 0 & 0 & \\ 0 & 0 & -\lambda_3 & 0 & \underline{0}_{4\times 3} \\ 0 & 0 & 0 & -\lambda_4 & \\ \hline 1 & 0 & 0 & -1 & \\ 0 & 1 & 0 & -1 & \underline{0}_{3\times 3} \\ 0 & 0 & 1 & -1 & \end{array}\right]$$

$$\underline{C} = \left[\begin{array}{c|ccc} \underline{0}_{3\times 4} & & \underline{I}_{3\times 3} & \\ \hline & 1 & -1 & 0 \\ \underline{0}_{3\times 4} & 1 & 0 & -1 \\ & 0 & 1 & -1 \end{array}\right]$$

$$\underline{B}_1 = \left[\begin{array}{cccccc} \frac{1}{M_1} & 0 & 0 & \frac{1}{M_1} & \frac{1}{M_1} & 0 \\ 0 & \frac{1}{M_1} & 0 & \frac{-1}{M_2} & 0 & \frac{1}{M_2} \\ 0 & 0 & \frac{1}{M_3} & 0 & \frac{-1}{M_3} & \frac{-1}{M_3} \\ \frac{-1}{M_4} & \frac{-1}{M_4} & \frac{-1}{M_4} & 0 & 0 & 0 \\ \hline & & & \underline{0}_{3\times 6} & & \end{array}\right]$$

$$\underline{B}_2 = \left[\begin{array}{cccccc} \frac{1}{M_1} & 0 & 0 & \frac{1}{M_1} & \frac{1}{M_1} & 0 \\ 0 & \frac{1}{M_2} & 0 & \frac{1}{M_2} & 0 & \frac{1}{M_2} \\ 0 & 0 & \frac{1}{M_3} & 0 & \frac{1}{M_3} & \frac{1}{M_3} \\ \frac{1}{M_4} & \frac{1}{M_4} & \frac{1}{M_4} & 0 & 0 & 0 \\ \hline & & \underline{0}_{3\times 6} & & & \end{array}\right]$$

$$f_1(\sigma_1) = E_1E_4B_{14}[\sin(\sigma_1+\delta_{14}^s) - \sin\delta_{14}^s],$$

$$g_1(\sigma_1) = E_1E_4G_{14}[\cos(\sigma_1+\delta_{14}^s) - \cos\delta_{14}^s]$$

$$f_2(\sigma_2) = E_2E_4B_{24}[\sin(\sigma_2+\delta_{24}^s) - \sin\delta_{24}^s],$$

$$g_2(\sigma_2) = E_2E_4G_{24}[\cos(\sigma_2+\delta_{24}^s) - \cos\delta_{24}^s]$$

$$f_3(\sigma_3) = E_3E_4B_{34}[\sin(\sigma_3+\delta_{34}^s) - \sin\delta_{34}^s],$$

$$g_3(\sigma_3) = E_3E_4G_{34}[\cos(\sigma_3+\delta_{34}^s) - \cos\delta_{34}^s]$$

$$f_4(\sigma_4) = E_1E_2B_{12}[\sin(\sigma_4+\delta_{12}^s) - \sin\delta_{12}^s],$$

$$g_4(\sigma_4) = E_1E_2G_{12}[\cos(\sigma_4+\delta_{12}^s) - \cos\delta_{12}^s]$$

$$f_5(\sigma_5) = E_1E_3B_{13}[\sin(\sigma_5+\delta_{13}^s) - \sin\delta_{13}^s],$$

$$g_5(\sigma_5) = E_1E_3G_{13}[\cos(\sigma_5+\delta_{13}^s) - \cos\delta_{13}^s]$$

$$f_6(\sigma_6) = E_2E_3B_{23}[\sin(\sigma_6+\delta_{23}^s) - \sin\delta_{23}^s],$$

$$g_6(\sigma_6) = E_2E_3G_{23}[\cos(\sigma_6+\delta_{23}^s) - \cos\delta_{23}^s]$$

n-machine System

For the general n machine case, the state variables are defined as

$$x_i = \omega_i \qquad i=1,2,\ldots,n$$

$$x_{n+i} = (\delta_i-\delta_n) - (\delta_i^s-\delta_n^s) \qquad i=1,2,\ldots,n-1$$

Define

$$\begin{aligned} f_i(\sigma_i) &= E_pE_qB_{pq}[\sin(\sigma_i+\delta_{pq}^s) - \sin\delta_{pq}^s] \\ g_i(\sigma_i) &= E_pE_qG_{pq}[\cos(\sigma_i+\delta_{pq}^s) - \cos\delta_{pq}^s] \end{aligned} \qquad (3.44)$$

For i=1,2,...,n-1

$$\sigma_i = (\delta_p-\delta_n) - (\delta_p^s-\delta_n^s) \qquad p=1,2,\ldots,n-1$$

For i > n-1

$$\sigma_i = (\delta_p-\delta_q) - (\delta_p^s-\delta_q^s) \qquad \begin{array}{l} p=1,2,\ldots,n-2 \\ q=2,3,\ldots,n-1 \\ (q > p) \end{array}$$

With these definitions, the state space model becomes

$$\begin{aligned} \dot{\underline{x}} &= \underline{A}\,\underline{x} - \underline{B}_1\underline{f}(\underline{\sigma}) - \underline{B}_2\underline{g}(\underline{\sigma}) \\ \underline{\sigma} &= \underline{C}\,\underline{x} \end{aligned} \qquad (3.45)$$

where the matrices $\underline{A}$, $\underline{B}_1$, $\underline{B}_2$ and $\underline{C}$ are given by

$$\underline{A} = \begin{array}{c} \begin{array}{cc} n & n-1 \end{array} \\ \left[\begin{array}{c|c} -\underline{D}\,\underline{M}^{-1} & \underline{0} \\ \hline \underline{K}_1^T & \underline{0} \end{array}\right] \end{array} \begin{array}{l} n \\ n-1 \end{array} \qquad \underline{B} = \begin{array}{c} m \\ \left[\begin{array}{c} \underline{M}^{-1}\underline{K} \\ \hline \underline{0} \end{array}\right] \end{array} \begin{array}{l} n \\ n-1 \end{array}$$

$$\underline{B}_2 = \begin{array}{c} m \\ \left[\begin{array}{c} \underline{M}^{-1}\underline{L} \\ \hline \underline{0} \end{array}\right] \end{array} \begin{array}{l} n \\ n-1 \end{array}$$

$$\underline{C} = \overset{n}{[\ \underline{0}}\quad \overset{n-1}{\underline{S}\ ]}\ m$$

$$= \begin{array}{c} \begin{array}{cc} n & n-1 \end{array} \\ \left[\begin{array}{c|c} \underline{0} & \underline{I} \\ \hline \underline{0} & \underline{J}^T \end{array}\right] \begin{array}{l} n-1 \\ m-n+1 \end{array} \end{array}$$

$$\underline{K} = \overset{\begin{array}{cccc} n-1 & n-2 & n-3 & 1 \end{array}}{\left[\begin{array}{cccc|cccc|cccc|c|c} 1 & 0 & \ldots & 0 & 1 & 1 & \ldots & 1 & 0 & \ldots & & 0 & & 0 \\ 0 & 1 & & 0 & -1 & 0 & \ldots & 0 & 1 & \ldots & & 1 & & \vdots \\ . & . & & . & 0 & -1 & \ldots & 0 & -1 & \ldots & & 0 & \ldots & 1 \\ . & . & & 1 & & & & -1 & & \ldots & & -1 & & -1 \\ -1 & -1 & \ldots & -1 & 0 & \ldots & \ldots & 0 & 0 & \ldots & & 0 & & 0 \end{array}\right]} n$$

$$= [\ \underline{K}_1 \mid \underline{K}_2 \mid \ldots\ \underline{K}_{n-1}\ ]$$

$$= \left[\begin{array}{c|c} \underline{K}_1 & \begin{array}{c} \underline{J} \\ \hline 0 \end{array} \end{array}\right]$$

where $\underline{I}$ = Identity matrix, $\underline{L}$ has same structure as $\underline{K}$ with -1 replaced by +1. It may be verified that

$$n\quad \underline{K}_1\ \overset{\begin{array}{cc} n-1 & m-n+1 \end{array}}{[\ \underline{I} \mid \underline{J}\ ]} = \overset{m}{[\ \underline{K}\ ]}\ n$$

i.e. $\quad \underline{K}_1\underline{S}^T = \underline{K}$ (3.46)

If the transfer conductances are neglected, then $\underline{g}(\underline{\sigma}) = 0$ and the model reduces to (3.40) with $\underline{B}_1 = \underline{B}$.

Uniform and Zero Damping

Four Machine System

$$\underline{x} = [\omega_1-\omega_4,\ \omega_2-\omega_4,\ \omega_3-\omega_4 \mid \delta_{14}-\delta_{14}^s,\ \delta_{24}-\delta_{24}^s,\ \delta_{34}-\delta_{34}^s]^T$$

We set $\lambda_1 = \lambda_2 = \lambda_3 = \lambda_4 = \lambda$ in (3.43) and subtract the fourth equation from the first three equations to get

$$\dot{\underline{x}} = \underline{A}\,\underline{x} - \underline{B}_1\,\underline{f}(\underline{\sigma}) - \underline{B}_2\,\underline{g}(\underline{\sigma})$$
$$\underline{\sigma} = \underline{C}\,\underline{x} \tag{3.47}$$

$$\underline{A} = \left[\begin{array}{ccc|c} -\lambda & 0 & 0 & \\ 0 & -\lambda & 0 & \underline{0} \\ 0 & 0 & -\lambda & \\ \hline 1 & 0 & 0 & \\ 0 & 1 & 0 & \underline{0} \\ 0 & 0 & 1 & \end{array}\right] \qquad \underline{C} = \left[\begin{array}{c|ccc} \underline{0} & & \underline{I} & \\ \hline & 1 & -1 & 0 \\ \underline{0} & 1 & 0 & -1 \\ & 0 & 1 & -1 \end{array}\right]$$

$$\underline{B}_1 = \begin{bmatrix} \frac{1}{M_1}+\frac{1}{M_4} & \frac{1}{M_4} & \frac{1}{M_4} & \frac{1}{M_1} & \frac{1}{M_1} & 0 \\ \frac{1}{M_4} & \frac{1}{M_2}+\frac{1}{M_4} & \frac{1}{M_4} & \frac{-1}{M_2} & 0 & \frac{1}{M_2} \\ \frac{1}{M_4} & \frac{1}{M_4} & \frac{1}{M_3}+\frac{1}{M_4} & 0 & \frac{-1}{M_3} & \frac{-1}{M_3} \end{bmatrix}$$

$$\underline{B}_2 = \begin{bmatrix} \frac{1}{M_1}-\frac{1}{M_4} & \frac{-1}{M_4} & \frac{-1}{M_4} & \frac{1}{M_1} & \frac{1}{M_1} & 0 \\ \frac{-1}{M_4} & \frac{1}{M_2}-\frac{1}{M_4} & \frac{-1}{M_4} & \frac{1}{M_2} & 0 & \frac{1}{M_2} \\ \frac{-1}{M_4} & \frac{-1}{M_4} & \frac{1}{M_3}-\frac{1}{M_4} & 0 & \frac{1}{M_3} & \frac{1}{M_3} \end{bmatrix}$$

n machine System

The state variables are

$$\underline{x} = [\omega_1-\omega_n, \omega_2-\omega_n, \ldots \omega_{n-1}-\omega_n \,|\, \delta_{1n}-\delta^s_{1n}, \delta_{2n}-\delta^s_{2n}, \ldots \delta_{n-1}-\delta^s_{n-1,n}]^T$$

The state model is given by

$$\dot{\underline{x}} = \underline{A}\,\underline{x} - \underline{B}_1\,\underline{f}(\underline{\sigma}) - \underline{B}_2\,\underline{g}(\underline{\sigma})$$
$$\underline{\sigma} = \underline{C}\,\underline{x} \tag{3.48}$$

where

$$\underline{A} = \begin{matrix} & n-1 & n-1 \\ n-1 & -\underline{\lambda} & \underline{0} \\ n-1 & \underline{I} & \underline{0} \end{matrix} \qquad \underline{B}_1 = \begin{bmatrix} \underline{K}_1^T \underline{M}^{-1} \underline{K} \\ \underline{0} \end{bmatrix} \begin{matrix} n-1 \\ n-1 \end{matrix}$$

$$\underline{C} = \begin{matrix} n-1 & n-1 & \\ \underline{0} & \underline{I} & n-1 \\ \underline{0} & \underline{J}^T & m-n+1 \end{matrix} \qquad \underline{B}_2 = \begin{bmatrix} \underline{K}_1^T \underline{M}^{-1} \underline{K} \\ \underline{0} \end{bmatrix} \begin{matrix} n-1 \\ n-1 \end{matrix}$$

For the zero damping case $\underline{\lambda} \equiv \underline{0}$ in (3.48).

3.7 State Space Models with SEP Transferred to the Origin (COA Reference Frame)

For the multi-machine system with the classical model, we finally consider the state space model in COA reference frame with SEP transferred to the origin. The procedure is analogous to that in Sec. 3.6. The state space models developed in Sec. 3.5 are to be converted so that origin is the equilibrium point.

Non-uniform Damping

In the case of non-uniform damping, Eqs. (3.24) and (3.25) are to be considered. These two sets of equations are coupled because of the damping terms and, hence, they cannot be cast in a state space form in any meaningful manner.

Uniform Damping

The equations to be considered are (3.27) and (3.29) which are repeated below for convenience.

$$\ddot{\delta}_o = -\frac{D_T}{M_T}\dot{\delta}_o + \frac{1}{M_T}P_{COA} \tag{3.49}$$

$$\ddot{\theta}_i = -\lambda\dot{\theta}_i + \frac{P_i}{M_i} - \frac{P_{ei}}{M_i} - \frac{P_{COA}}{M_T} \qquad i=1,2,\ldots,n \tag{3.50}$$

The post fault stable equilibrium point $\underline{\theta}^s$ is obtained from (3.50) by setting $\theta_i = \dot{\theta}_i = 0$, i.e.

$$P_i - P_{ei}(\underline{\theta}^s) - \frac{M_i}{M_T} P_{COA}(\underline{\theta}^s) = 0$$

The minimal dimension of the state space is 2n-2 and we accordingly define the state variables as

$$\underline{x}^T = [\tilde{\omega}_1-\tilde{\omega}_n, \tilde{\omega}_2-\tilde{\omega}_n, \ldots \tilde{\omega}_{n-1}-\tilde{\omega}_n \mid \theta_{1n}-\theta_{1n}^s, \theta_{2n}-\theta_{2n}^s, \ldots \theta_{n-1,n}-\theta_{n-1,n}^s]$$

where

$$\theta_{in} = \theta_i - \theta_n, \quad \tilde{\omega}_i = \dot{\theta}_i.$$

The state equations are obtained by first subtracting the nth equation in (3.50) from the other equations and then expressing the resulting equations as

$$\dot{\tilde{\omega}}_i - \dot{\tilde{\omega}}_n = -\lambda(\tilde{\omega}_i - \tilde{\omega}_n) + \left(\frac{P_{ei}(\underline{\theta}^s)}{M_i} - \frac{P_{en}(\underline{\theta}^s)}{M_n}\right) - \left(\frac{P_{ei}(\underline{\theta})}{M_i} - \frac{P_{en}(\underline{\theta})}{M_n}\right)$$

and

$$\dot{\theta}_{in} - \dot{\theta}_{in}^s = \tilde{\omega}_i - \tilde{\omega}_n \qquad i=1,\ldots,n-1 \qquad (3.51)$$

Since $\theta_{in} = \delta_{in}$ and $\omega_{in} = \tilde{\omega}_{in}$, it will turn out that Eq. (3.51) is identical to (3.48). For zero damping case $\lambda \equiv 0$.

We, therefore, conclude that for uniform and zero damping case, if we cast the equations in minimal state space, it is immaterial whether we take machine angle or center of angle as the reference.

3.8 State Space Models Including Flux Decay

One of the more realistic efforts at improving the synchronous machine model is not to assume the voltage behind transient reactance as constant. This would then account partially for the flux decay effects [13]. In the model which includes flux decay effects, we shall assume that $|E_i|$ (i=1,2,...,n) in (3.12) and (3.13) are no longer held constant. We will, however, assume that the transfer conductances are negligible and $|E_i|$ is held constant only in the expression for P_i. Hence, Eq. (3.37) applies except that $|E_i|, |E_j|$ are now time varying. Let the values of $|E_i|, |E_j|$ in the post-fault steady state be $|E_i|^s$ and $|E_j|^s$. Then Eq. (3.37) becomes (dropping the $|\cdot|$ notation

$$M_i \frac{d^2\delta_i}{dt^2} + D_i \frac{d\delta_i}{dt} = \sum_{\substack{j=1 \\ \neq i}}^{n} B_{ij}(E_i^s E_j^s \sin\delta_{ij}^s - E_i E_j \sin\delta_{ij}) \qquad i=1,2,\ldots,n \tag{3.52}$$

The change of the internal voltage of ith generator is described by the first order differential equation

$$\frac{dE_i}{dt} = -\alpha_i(E_i - E_i^s) - \beta_i \sum_{j=1}^{n} B_{ij}E_j(\cos\delta_{ij}^s - \cos\delta_{ij}) \qquad \text{for } i=1,2,\ldots,n \tag{3.53}$$

where α_i and β_i are constants (see Ref. [13] for details).

For the system comprising of (3.52) and (3.53), the minimal order of the state space is 2n-1+n, i.e. 3n-1. The 3n-1 state variables are defined by

$$\underline{x} = [\omega_1,\ldots,\omega_n \,\vdots\, \delta_{1n}-\delta_{1n}^s,\ldots,\delta_{n-1,n}-\delta_{n-1,n}^s$$

$$E_1-E_1^s,\ldots,E_n-E_n^s]^T$$

$$= [x_1,\ldots,x_n \,\vdots\, x_{n+1},\ldots,x_{2n-1} \,\vdots\, x_{2n},\ldots,x_{3n-1}]^T$$

The state equations are

$$\begin{aligned} \dot{\underline{x}} &= \underline{A}\,\underline{x} - \underline{B}\,\underline{F}(\underline{\sigma}) \\ \underline{\sigma} &= \underline{C}\,\underline{x} \end{aligned} \tag{3.54}$$

$$
\underline{A} = \begin{array}{c} \quad n \qquad n-1 \quad n \\ \left[\begin{array}{c|c|c} -\underline{DM}^{-1} & \underline{0} & \underline{0} \\ \hline \underline{K}_1^T & \underline{0} & \underline{0} \\ \hline \underline{0} & \underline{0} & -\alpha \end{array}\right] \begin{array}{c} n \\ n-1 \\ n \end{array} \end{array} \quad
\underline{B} = \begin{array}{c} \quad m \qquad n \\ \left[\begin{array}{c|c} \underline{M}^{-1}\underline{K} & \underline{0} \\ \hline \underline{0} & \underline{0} \\ \hline \underline{0} & \beta \end{array}\right] \begin{array}{c} n \\ n-1 \\ n \end{array} \end{array}
$$

where $\underline{A}$ is $(3n-1)\times(3n-1)$ matrix, $\underline{B}$ is $(3n-1)\times(m+n)$ and $\underline{C}$ is $(m+n)\times(3n-1)$ matrix, $\underline{K}_1$ and $\underline{K}$ are as defined in Eq. (3.37), $\underline{\alpha}$ and $\underline{\beta}$ are diagonal $n\times n$ matrices.

$$
\underline{C} = \begin{array}{c} \quad n \qquad n-1 \quad n \\ \left[\begin{array}{c|c|c} \underline{0} & \underline{I} & \underline{0} \\ \hline \underline{0} & \underline{J}^T & \underline{0} \\ \hline \underline{0} & \underline{0} & \underline{I} \end{array}\right] \begin{array}{c} n-1 \\ (m-n+1) \\ n \end{array} \end{array} \qquad
\underline{F}(\underline{\sigma}) = \left[\begin{array}{c} \underline{f}_1(\underline{\sigma}) \\ \hline \underline{f}_2(\underline{\sigma}) \end{array}\right] \begin{array}{c} m \\ n \end{array}
$$

$$
f_{1i}(\underline{\sigma}) = B_{pq}[E_p E_q \sin(\sigma_i + \delta_{pq}^S) - \sin\delta_{pq}^S] \qquad i=1,2,\ldots,m
$$

$$
f_{2i}(\underline{\sigma}) = \sum_{j=1}^{n} B_{ij}E_j(\cos\delta_{ij}^S - \cos\delta_{ij}) \qquad i=1,2,\ldots,n
$$

The above system describes a nonlinear system with multiple, memoryless, coupled nonlinearities. They do not satisfy the sector condition just as systems with transfer conductances do not. However, by putting them in a Luré type form and using a modified form of the multi-variable Popov criterion [14] constructive procedure is possible to arrive at the Lyapunov function.

Conclusion

In this chapter, we have discussed several state space descriptions of an n-machine system with non-uniform, as well as uniform damping with zero as well as non-zero transfer conductances and finally a model including flux decay effects.

REFERENCES

1. Anderson, P. M. and Fouad, A. A., "Power System Control and Stability", (Book), Iowa State University Press, Ames, Iowa, 1977.
2. Stagg, G. W. and El-Abiad, A. H., "Computer Methods in Power System Analysis", (Book), McGraw-Hill, New York, 1968.
3. Pai, M. A., "Computer Techniques in Power System Analysis", (Book), Tata McGraw Hill (India), New Delhi, 1979.
4. Willems, J. L., "A Partial Stability Approach to the Problem of Transient Power System Stability", International Journal of Control, Vol. 19, No. 1, 1974, pp. 1-14.
5. Sastry, V. R. and Murthy, P. G., "Derivation of Completely Controllable and Completely Observable State Models for Multi-machine Power System Stability Studies", International Journal of Control, Vol. 16, No. 4, October 1972, pp. 777-788.
6. Sastry, V. R. and Murthy, P. G., Discussion of Willems' paper, "Direct Methods of Transient Stability Studies in Power Systems Analysis", and reply by author IEEE Trans. Automatic Control, Vol. AC-17, No. 4, August 1972, pp. 480-581.
7. Pai, M. A. and Murthy, P. G., "New Lyapunov Functions for Power Systems Based on Minimal Realizations", International Journal of Control, Vol. 19, No. 2, February 1974, pp. 401-415.
8. Tavora, G. J. and Smith, O.J.M., "Stability Analysis of Power Systems", IEEE Trans. on Power Apparatus and Systems, Vol. PAS 91, No. 3, May/June 1971, pp. 1138-1139.
9. Stanton, K. N., "Dynamic Energy Balance Studies for Simulation of Power Frequency Transients", Proc. PICA Conference, Boston, May 1971.
10. Fouad, A. A. and Lugtu, R. L., "Transient Stability Analysis of Power Systems Using Lyapunov's Second Method", IEEE Winter Power Meeting, Paper No. C72, 145-6, New York, Feb. 1972.
11. Padiyar, K. R. and Doreswamy, C. V., "Analysis of Dynamic Equilibrium of Power Systems", Proc. 7th Annual South Eastern Symposium on System Theory, March 20-21, 1975, pp. 123-125.
12. Gilbert, E. G., "Controllability and Observability in Multivariable Control Systems", SIAM Journal (Control), Vol. 1, 1962-63, pp. 128-151.

13. Kakimoto, N., Ohsawa, Y. and Hayashi, M., "Transient Stability Analysis of Multimachine Power Systems with Field Flux Decays via Lyapunov's Direct Method", IEEE Trans. PAS, Vol. 99, No. 6, Sept/Oct. 1980, pp. 1819-1827.

14. Desoer, C. A., Wu, M. Y., "Stability of a Nonlinear Time Invariant Feedback System Under Almost Constant Inputs", Automatica, Vol. 5, 1969, pp. 231-233.

Chapter IV

LYAPUNOV FUNCTIONS FOR POWER SYSTEMS

4.1 Introduction

In the previous chapter, we developed the mathematical models of multimachine power systems used for transient stability analysis. In Chapter I (sec. 1.2), it was shown that the critical clearing time t_{cr} can be computed from knowing the region of attraction around the post-fault stable equilibrium point. This is accomplished by constructing a Lyapunov function V(x) for the post-fault system and defining a region Ω by the inequality $V(\underline{x}) < C$ so that inside $\Omega, V(\underline{x}) > 0$ and $\dot{V}(\underline{x}) < 0$. If $\dot{V}(\underline{x}) = 0$ inside Ω, then the points $\dot{V}(\underline{x}) = 0$ should not constitute a trajectory of the system excepting $\underline{x} = 0$. The accuracy to which t_{cr} approximates the actual clearing time depends both on the quality of $V(\underline{x})$ and the value of C which defines Ω. In general due to the sufficiency nature of Lyapunov's theorems, the region defined by $V(\underline{x}) < C$ is contained in the exact region of stability. In this chapter, we shall discuss the systematic procedures to construct $V(\underline{x})$. This is done for both single and multimachine systems using the different techniques explained in Chapter II. While for the simple models, different methods may yield the same $V(\underline{x})$, the methods will be useful when complex models are considered. Techniques for computing the region of attraction are discussed in Chapter VI.

4.2 Single Machine System

Consider a single machine infinite bus case whose swing equation is given by

$$M \frac{d^2\delta}{dt^2} + D \frac{d\delta}{dt} = P - P_e \sin \delta \tag{4.1}$$

where $P_e \sin \delta = \frac{E_1 E_2}{X_{eq}} \sin \delta$ is the electrical power delivered and $P = P_m - E_1^2 G_{11}$. P_m is the mechanical power input, E_1 is the internal voltage magnitude of the machine, E_2 is the magnitude of the voltage of the infinite bus and G_{11} is the self-conductance at the internal bus after eliminating the generator bus where the load, if any, is converted into a constant admittance. X_{eq} is the reactance corresponding to the post-fault system between the generator internal node and the infinite bus. Equation (4.1) will have the equilibrium points given by the solution of the nonlinear algebraic equation

$$P - P_e \sin \delta = 0 \tag{4.2}$$

The solutions of this equation in an interval of $-\pi$ to $+\pi$ are

(i) $\delta = \delta^s = \sin^{-1} \frac{P}{P_e}$

(ii) $\delta = \delta^u = \pi - \sin^{-1} \frac{P}{P_e} = \pi - \delta^s$

It is easily verified by linearization of (4.1) that δ^s is the stable and δ^u is the unstable equilibrium point. In fact, the unstable equilibrium point can be shown to be a saddle point.

For Lyapunov stability analysis, it is convenient to transfer the post-fault stable equilibrium point (SEP) to the origin by the transformation $x = \delta - \delta^s$. Thus, (4.1) becomes

$$M \frac{d^2x}{dt^2} + D \frac{dx}{dt} = P_e \sin \delta^s - P_e \sin (x+\delta^s)$$

$$= -(P_e \sin (x+\delta^s) - P_e \sin \delta^s) \tag{4.3}$$

Equation (4.3) is case in the state variable form by defining $x_1 = x = \delta - \delta^s$, $x_2 = \dot{x} = \dot{\delta} = \omega$. Thus

$$\dot{x}_1 = x_2 \qquad \dot{x}_2 = -\frac{D}{M} x_2 - \frac{1}{M} f(x_1) \tag{4.4}$$

where $f(x_1) = P_e \sin(x_1+\delta^s) - P_e \sin \delta^s$. Equation (4.4) has the origin $x_1 = x_2 = 0$ as the stable equilibrium point. The nonlinearity $f(x_1)$ has an interesting property, namely that it passes through the origin and is in the first and third quadrant in a region around the origin (Fig. 4.1). The nonlinearity $f(x_1)$ is the power angle curve $P_e \sin \delta$ shifted down by an amount $P = P_e \sin \delta^s$ and with the origin shifted to $\delta = \delta^s$ (Fig. 4.2). It is easily verified that the intersection of the horizontal line of constant P in Fig. 4.2 occurs at δ^s, $\pi - \delta^s$, $-\pi - \delta^s$. This graphical picture is useful in multi-machine systems when we discuss the unstable equilibrium points.

4.3 Lyapunov's Method and Equal Area Criterion (Method of First Integrals)

In this section, we provide a link up of Lyapunov's method with the classical equal area criterion which is generally discussed for the zero damping case. We apply the method of first integrals discussed in Sec. 2.11. Consider the swing equation (4.1) with D = 0

$$M \frac{d^2\delta}{dt^2} = P - P_e \sin \delta \tag{4.5}$$

The state equations are

$$\begin{aligned} \dot{x}_1 &= x_2 \\ \dot{x}_2 &= -\frac{1}{M} f(x_1) \end{aligned} \tag{4.6}$$

Since $\frac{\partial f_1}{\partial x_1} + \frac{\partial f_2}{\partial x_2} = 0$, Eq. (4.6) has a first integral $V(x_1,x_2)=C$. From (4.6) we obtain

$$\frac{dx_1}{dx_2} = \frac{x_2}{(-1/M)f(x_1)} \tag{4.7}$$

The first integral is obtained as

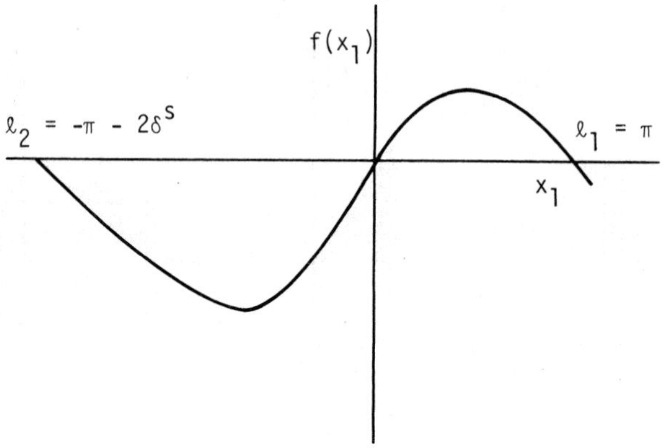

Fig. 4.1. The nonlinearity $f(x_1)$.

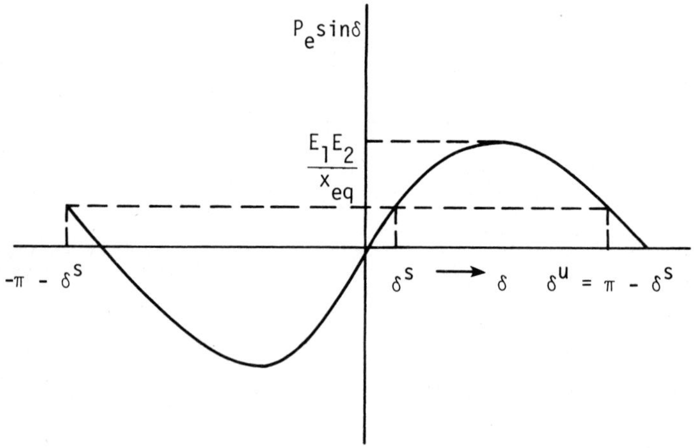

Fig. 4.2. Illustrating the power angle nonlinearity.

$$V(x_1,x_2) = \frac{x_2^2}{2} + \frac{1}{M}\int_0^{x_1} f(u)du$$

$$= \frac{x_2^2}{2} + \frac{1}{M}\int_0^{x_1} (P_e \sin(u+\delta^s) - P_e \sin\delta^s)du \quad (4.8)$$

Multiplying the right-hand side of (4.8) by M and evaluating the integral, we get

$$V(x_1,x_2) = \frac{1}{2} Mx_2^2 - P_e \cos(x_1+\delta^s)$$
$$+ P_e \cos\delta^s - x_1 P_e \sin\delta^s \quad (4.9)$$

It is easily verified that $\dot{V}(x_1,x_2) = 0$. Since $V > 0$ and $\dot{V} = 0$, the origin is stable in the sense of Lyapunov. To relate V to the physical variables, substitute $x_1 = \delta - \delta^s$ and $x_2 = \dot{\delta} = \omega$ in Eq. (4.9) which now becomes

$$V(\delta,\omega) = \frac{1}{2} M\omega^2 - P_e \cos\delta + P_e \cos\delta^s$$
$$- (\delta-\delta^s)P_e \sin\delta^s \quad (4.10)$$

Since $P = P_e \sin\delta^s$

$$V(\delta,\omega) = \frac{1}{2} M\omega^2 - P_e(\cos\delta-\cos\delta^s) - P(\delta-\delta^s) \quad (4.11)$$

The constant V-contours in the state space are shown in Fig. 4.3. Since $V(\delta,\omega)$ is also a first integral of motion, the constant V-contours are also the phase plane trajectories. Physically $\frac{1}{2} M\omega^2 = V_k \triangleq$ Kinetic Energy and $-[P(\delta-\delta^s) + P_e(\cos\delta-\cos\delta^s)] \triangleq$ Potential Energy V_p. Potential energy can be further interpreted as the sum of positional energy $-P(\delta-\delta^s)$ of the rotor and magnetic stored energy in the reactance x_{eq} $-P_e(\cos\delta-\cos\delta^s)$. Since the system is conservative,

$$V(\delta,\omega) = V_k(\omega) + V_p(\delta) = \text{constant} \quad (4.12)$$

The separatrix passing through the saddle point $(\delta^u,0)$ is defined by the constant V-curve $V(\delta,\omega) = V(\delta^u,0)$. The closed region inside of the separatrix defines a region of stability.

From Fig. (4.4) and Eq. (4.11), $V(\delta^u,0) = -P(\delta^u-\delta^s) - P_e(\cos\delta^u-\cos\delta^s)$ = Area 2 + Area 3. During the faulted period $M \frac{d^2\delta}{dt^2} = P - P_e^f \sin\delta$ where $P_e^f = \frac{E_1E_2}{x_f}$, x_f = reactance between internal node and infinite bus during the faulted period.

$$\text{Area 1} = \int_{\delta^o}^{\delta} (P - P_e^f \sin\delta)d\delta = \int_{\delta^o}^{\delta} M \frac{d\omega}{dt} d\delta$$

$$= \int_{\omega^o}^{\omega} M\omega \, d\omega = \frac{1}{2} M\omega^2$$

since $\omega^o = 0$ at the pre-fault operating point. Hence, from (4.11) $V(\delta,\omega)$ = Area 1 + Area 3. The region of stability is defined by

$$V(\delta,\omega) < V(\delta^u,0) \tag{4.13}$$

i.e. Area 1 + Area 3 < Area 2 + Area 3

i.e. Area 1 < Area 2

This is precisely the equal area criterion in the literature. With damping, the equal area criterion gets modified as discussed in Ref. [1].

4.4 Variable Gradient Method

We apply the method (sec. 2.13 of Chapter II) to Eq. (4.4). Assume a gradient vector of the following form

$$\nabla \underline{V} = \begin{bmatrix} \nabla V_1 \\ \nabla V_2 \end{bmatrix} = \begin{bmatrix} \alpha_{11}x_1 + \alpha_{12}x_2 \\ \alpha_{21}x_1 + \alpha_{22}x_2 \end{bmatrix} \tag{4.14}$$

The curl condition is $\frac{\partial \nabla V_1}{\partial x_2} = \frac{\partial \nabla V_2}{\partial x_1}$. $\dot{V}$ is computed as

$$\dot{V} = x_1x_2(\alpha_{11}-\alpha_{21}\ D/M) + x_2^2(\alpha_{12}-\alpha_{22}\ D/M)$$
$$- (\alpha_{21}x_1+\alpha_{22}x_2)\frac{1}{M} f(x_1) \tag{4.15}$$

The simplest way of constraining $\dot{V} \leq 0$ is to start by setting $\alpha_{12} = 0$. The curl equation is

$$\alpha_{12} = \alpha_{21} = \alpha_{21,k} + x_1 \frac{\partial \alpha_{21,v}}{\partial x_1} \qquad (4.16)$$

Since $\alpha_{12} = 0$, we have $\alpha_{21} = 0$ also. Letting $\alpha_{11} = \frac{\alpha_{22}}{M} \frac{f(x_1)}{x_1}$ and $\alpha_{22} = 1$ we get

$$\dot{V} = (-D/M)\, x_2^2 \qquad (4.17)$$

and

$$\nabla \underline{V} = \begin{bmatrix} \frac{1}{M} f(x_1) \\ x_2 \end{bmatrix}$$

Using (2.16) to integrate $\nabla \underline{V}$, we obtain $V(x_1,x_2)$ as

$$V(x_1,x_2) = \frac{x_2^2}{2} + \frac{1}{M} \int_0^{x_1} f(u)du) \qquad (4.18)$$

Multiplying the right-hand side of (4.18) by M gives the same Lyapunov function as for the conservative system. The damping term however does not appear in the V-function. To do this, we can allow $\dot{V}$ to be negative semi-definite function of x_1, by choosing $\alpha_{22} = 2$ and $\alpha_{12} = \alpha_{21} = 2\frac{D}{M}$. Then

$$\dot{V} = x_1 x_2 (\alpha_{11} - 2\frac{D^2}{M^2}) - 2(\frac{D}{M} x_1 + x_2)\frac{f(x_1)}{M} \qquad (4.19)$$

Suppose we further set $\alpha_{11} = 2(\frac{D^2}{M^2} x_1 + \frac{f(x_1)}{M})/x_1$ then, $\dot{V} = -2\frac{D}{M} x_1 \frac{f(x_1)}{M}$ which is negative semi-definite in a region around the origin.

$$\nabla \underline{V} = \begin{bmatrix} 2(\frac{D^2}{M^2} x_1 + \frac{f(x_1)}{M}) + \frac{2D}{M} x_2 \\ \frac{2D}{M} x_1 + 2x_2 \end{bmatrix} \qquad (4.20)$$

Integration of $\nabla \underline{V}$ gives

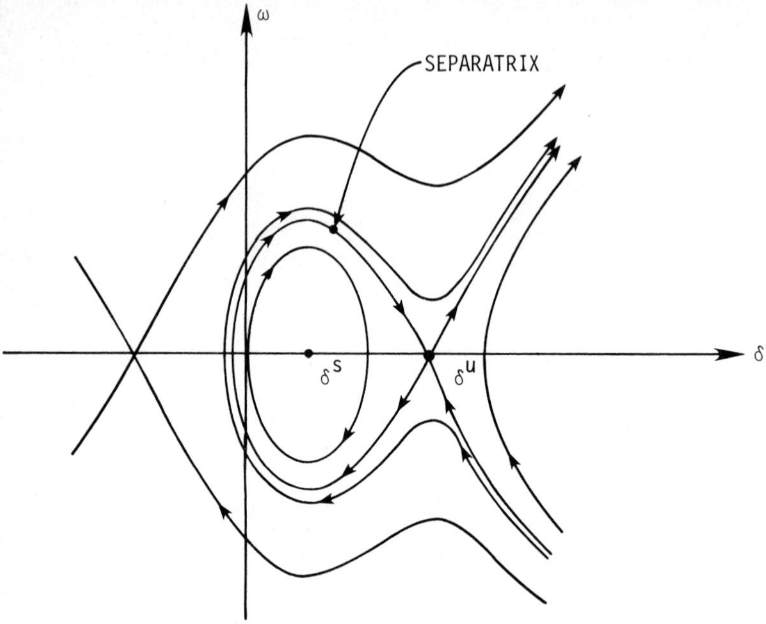

Fig. 4.3. Constant V contours and phase trajectories in δ-ω plane.

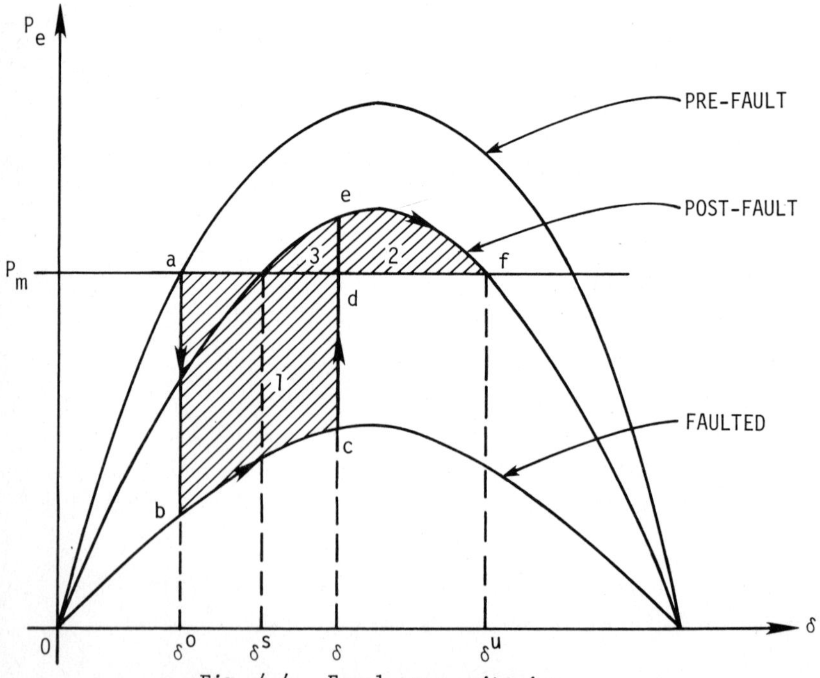

Fig. 4.4. Equal area criterion.

$$V(x_1,x_2) = \int_0^{x_1} \nabla V_1(\gamma_1,0)d\gamma_1 + \int_0^{x_2} \nabla V_2(x_1,\gamma_2)d\gamma_2$$

$$= \frac{D^2}{M^2} x_1^2 + \frac{2}{M}\int_0^{x_1} f(u)du + \frac{2D}{M} x_1x_2 + x_2^2 \quad (4.21)$$

Multiplying the right-hand side of (4.21) by $\frac{M}{2}$ gives

$$V(x_1,x_2) = \frac{1}{2} M(\frac{Dx_1}{M} + x_2)^2 + \int_0^{x_1} f(u)du \quad (4.22)$$

If D = 0, the above Lyapunov function reduces to the energy function derived previously. The region of asymptotic stability is obtained as

$$V(x_1,x_2) < V(\delta^u,0)$$

where δ^u is the saddle point given by $\delta^u = \pi - \delta^s$.

For single machine systems, there have been other approaches to the construction of Lyapunov functions such as Cartwright and Aizerman's Method [2], format method [3], energy metric algorithm method [4], etc. All of these methods yield the same Lyapunov function as in the variable gradient method for the classical model. Mansour [4] has extended the energy metric algorithm method to the single machine infinite bus case when the machine is represented by a third order model including flux decay effects.

4.5 Zubov's Method [5]

This method is attractive for two reasons; (i) theoretically it gives the exact region of stability and (ii) it is suited for computer implementation if the nonlinearity is expressed as a power series. Although a decade ago, the computer memory requirements using this method was considered as a deterrent, it is no longer true now. The theory of Zubov's method has been discussed in Chapter II (Sec. 2.14).

We consider a slightly more complex model of the synchronous machine than in the previous section. Effect of transient saliency and nonlinear load damping is considered. The derivation of the model is contained in Ref. [5].

The swing equation is given by

$$M \frac{d^2\delta}{dt^2} + D(\delta) \frac{d\delta}{dt} = P - P_e \sin\delta - P_s \sin 2\delta$$
$$D(\delta) = a_1 \sin^2\delta + a_2 \cos^2\delta \tag{4.23}$$

where M, P, P_e, P_s, a_1, a_2 are all functions of system and machine parameters. The equation is normalized by defining $\tau = t\sqrt{P_e/M}$ which yields

$$\frac{d^2\delta}{d\tau^2} + D'(\delta) \frac{d\delta}{dt} = P' - \sin\delta - P'_s \sin 2\delta \tag{4.24}$$

where

$$D'(\delta) = D(\delta)\sqrt{\frac{1}{P_e M}}, \quad P'_s = \frac{P_s}{P_e}, \quad P' = \frac{P}{P_e}$$

Assuming that the equation pertains to the post-fault state, the stable equilibrium point (SEP) δ^s is obtained by solving

$$\sin\delta + P'_s \sin 2\delta = P'$$

Defining the transformation $x = \delta - \delta^s$ which transfers the SEP to the origin, Eq. (4.24) becomes

$$\frac{d^2x}{d\tau^2} + D'(x) \frac{dx}{d\tau} + R(x) = 0 \tag{4.25}$$

$D'(x)$ and $R(x)$ composed of sine and cosine terms can be expanded in a power series as

$$D'(x) = D_0 + D_1 x + D_2 x^2 + \ldots$$
$$R(x) = R_1 x + R_2 x^2 + \ldots \tag{4.26}$$

Defining state variables as $x_1 = x$, $x_2 = \dot{x}$, Eq. (4.25) can be written as

$$\dot{x}_1 = x_2$$
$$\dot{x}_2 = -D'(x_1)x_2 - R(x_1) \quad (4.27)$$

Following the method discussed in Sec. 2.14 of Chapter II, the Lyapunov function is chosen as

$$V^{(n)}(x_1,x_2) = V_2(x_1,x_2) + V_3(x_1,x_2) \ldots + V_n(x_1,x_2)$$
$$= \sum_{j=2}^{n} \sum_{k=0}^{j} d_{jk} x_1^{j-k} x_2^{k} \quad (4.28)$$

where $V_m(x_1,x_2)$ is homogeneous of degree m in x_1,x_2.

i.e.
$$V_2(x_1,x_2) = d_{20}x_1^2 + d_{11}x_1x_2 + d_{02}x_2^2$$
$$V_3(x_1,x_2) = d_{30}x_1^3 + d_{21}x_1^2x_2 + d_{12}x_1x_2^2 + d_{03}x_3^3 \quad (4.29)$$

The d_{jk} coefficients are determined recursively as discussed in Sec. 2.14.3. $\phi(x_1,x_2)$ in the p.d.e. (2.24) is chosen as $\phi(x_1,x_2) = \alpha x_1^2 + \beta x_2^2$, $\alpha \geq 0$, $\beta \geq 0$, $\alpha + \beta \neq 0$.

4.5.1 Numerical Example [5]

A synchronous machine is connected to the transmission system as shown in Fig. 4.5. The complex power delivered by the machine is 0.753 + j0.03 p.u. and the initial terminal voltage of the machine is 1.05 p.u. A sudden 3-phase to ground fault is placed at the transmission line A-B near the bus A. Line A-B is cleared at 0.0833 sec. and after the fault is cleared, the line is reconnected to the system at 0.4 sec. Investigate if the system is stable. The following data and system differential equations pertaining to the system are provided: Voltage behind transient reactance = $1.055\angle 50.83°$, Infinite bus voltage = $1.058\angle 0°$, the dimensionless time τ = 7.308t.

The swing equation during the faulted state until t = 0.0833 sec. is

$$\frac{d^2x}{d\tau^2} = 0.6489 \tag{4.30}$$

The swing equation for $.0833 \leq t \leq .4$ is

$$\frac{d^2x}{d\tau^2} + (0.03072 \cos^2 x + 0.01332 \sin^2 x)\frac{dx}{d\tau} + 0.7258 \sin x$$
$$- 0.07227 \sin 2x = .6489 \tag{4.31}$$

The swing equation for $t \geq .4$ sec. is

$$\frac{d^2x}{d\tau^2} + (0.04972 \cos^2 x + 0.02468 \sin^2 x)\frac{dx}{d\tau}$$
$$+ \sin x - 0.1291 \sin 2x = 0.6489 \tag{4.32}$$

We first transform the post-fault equation (4.32) into the state space form as

$$\frac{dx_1}{d\tau} = x_2$$
$$\frac{dx_2}{d\tau} = -D'(x_1)x_2 - R(x_1) \tag{4.33}$$

The power series expansion of $D'(x_1)$ and $R(x_1)$ is obtained as

$$D'(x_1) = D_0 + D_1x_1 + \ldots.\ D_\ell x^\ell + \ldots.$$
$$= D_0 + \sum_{\ell=1}^{\infty} \frac{2^\ell}{\ell!}(0.1252)\cos(2\delta^s + \frac{\ell\pi}{2})x_1^\ell \tag{4.34}$$

where $D_0 = 0.0372 + 0.01252 \cos 2\delta^s$. $\delta^s = \delta^o = 50.83°$ since the final configuration is same as the initial configuration.

$$R(x_1) = \sum_{\ell=1}^{\infty} \frac{1}{\ell!}\{\cos(\delta^s + \frac{(\ell-1)\pi}{2})$$
$$- 2^\ell(0.1291)\cos(2\delta^s + \frac{(\ell-1)\pi}{2})\}x_1^\ell \tag{4.35}$$

In the computation it was found sufficient to truncate both $D'(x_1)$ and $R(x_1)$ at $\ell = 30$. The approximate stability regions computed from the truncated Lyapunov function $V^{(n)}$ for n = 5,8,10,16 and 26 are shown in Fig. 4.6. The function $\phi(x_1,x_2)$ chosen in the p.d.e. is $\phi = \alpha x_1^2 + \beta x_2^2$

$$\alpha = \beta = D_0 = 0.0372 + 0.0125 \cos 2\delta^s$$

The region of stability increases with n initially for n = 5,8,10 and 16 but decreases for n = 26 indicating that the approximate region of stability does not approach the true boundary monotonically with increase of n. The stability region for n = 35 was computed but indicated no improvement over that of n = 16. The best value of n depends on the particular example. This constitutes one of the drawbacks of this method. Figure 4.7 shows an example of how the boundary of a stability region $V^{(8)}(\underline{x})$ is determined. As explained in Sec. 2.14.2 of Chapter II, the d_{jk} coefficients are first determined. The points where $\dot{V}^{(8)}(\underline{x}) = 0$ are searched until minimum value of $V^{(8)}(\underline{x})$ is found. This minimum value denoted by C defines the stability region as $V^{(8)}(\underline{x}) < C$. Among the different choices of $\phi(x_1,x_2)$, the one with $\phi = D_0(x_1^2+x_2^2)$ gave the best results with C = 0.5057.

To ascertain the stability of the system, the faulted system equations (4.30) and (4.31) are integrated from the instant of fault until that of final switching. For a fault clearing at t = .0833 sec. and final switching at t = 0.4 sec., it is found that at t = 0.4 sec., the state of the system falls inside the stability region $V^{(10)}$, $V^{(16)}$, and $V^{(26)}$ although outside of $V^{(5)}$ and $V^{(8)}$. The power system is therefore stable since Lyapunov's method gives only sufficient conditions.

The extension of the above method for the case when flux decay effect is included is also possible. The differential equations now become

$$M \frac{d^2\delta}{dt^2} + D(\delta)\frac{d\delta}{dt} = P - P_e \sin\delta - P_s \sin 2\delta$$

$$\frac{dE_q'}{dt} = K E_{ex} - L E_q' + M \cos\delta \tag{4.36}$$

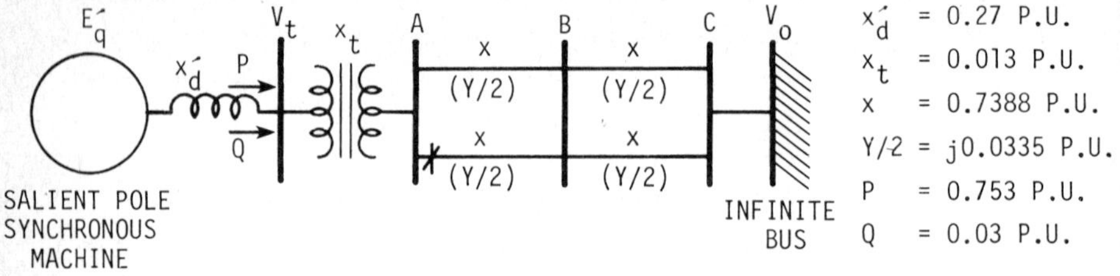

Fig. 4.5. Single machine infinite bus system.

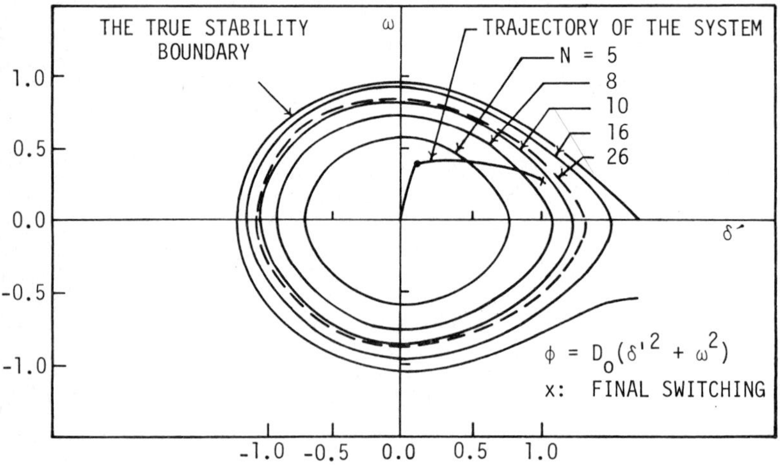

Fig. 4.6. Stability boundaries by Zubov's method for different values of n.

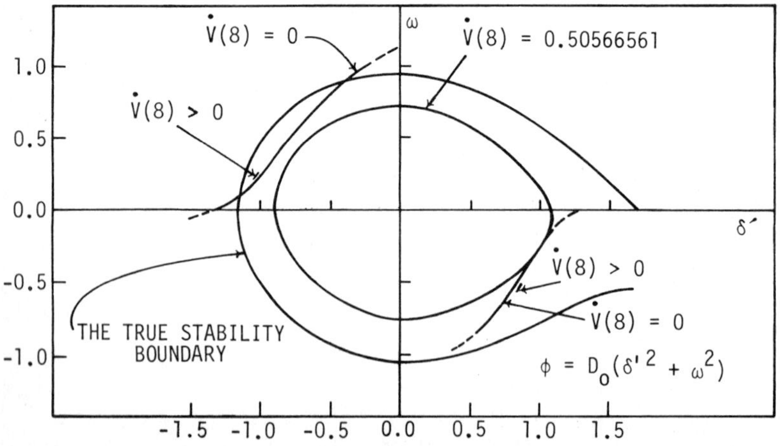

Fig. 4.7. Computation of stability boundary for n = 8.
(Reproduced from Ref. [5]).

where K, L and M are constants depending on machine and system parameters. Although the procedure is same as in the previous example, programming requirements for computer implementation are considerable. For details, refer to Ref. [6].

4.6 Popov's Method - Single Machine Case [7]

As discussed in Sec. 4.2, the nonlinearity for the single machine infinite bus case lies in the first and third quadrants in a region around the origin (Fig. 4.1). This implies two things: (i) we can apply Popov's theorem for absolute stability, (ii) the absolute stability property will be modified so as to have a region of attraction since the nonlinearity violates the sector condition for outside the interval $\ell_2 < \sigma < \ell_1$ ($\ell_1 > 0$ and $\ell_2 < 0$). In view of (i) we can construct systematically the Lyapunov function via Popov's theorem and Kalman's construction procedure as discussed in Sec. 2.15.4. It is also possible to get an analytical expression for an estimate of the region of attraction from a knowledge ℓ_1 and ℓ_2.

The state space equations (4.4) are rewritten in the form

$$\begin{aligned} \dot{\underline{x}} &= \underline{A}\underline{x} + \underline{b}\, f(\sigma) \\ \underline{\sigma} &= \underline{c}^T \underline{x} \end{aligned} \tag{4.37}$$

where $\underline{A} = \begin{bmatrix} 0 & 1 \\ 0 & -D/M \end{bmatrix}$ $\quad \underline{b} = \begin{bmatrix} 0 \\ -1/M \end{bmatrix}$ $\quad \underline{c} = \begin{bmatrix} 1 \\ 0 \end{bmatrix}$

$$f(\sigma) = P_e(\sin(\sigma+\delta^s) - \sin\delta^s) = P_e \sin(\sigma+\delta^s) - P \tag{4.38}$$

The transfer function of the linear part is given by (2.44) as $G(s) = \underline{c}^T(sI-\underline{A})^{-1}\underline{b}$ which yields

$$G(s) = \frac{1}{s(s+D/M)}$$

The nonlinearity (4.38) is shown in Fig. 4.1 and is seen to satisfy the sector condition $0 < \frac{f(\sigma)}{\sigma} < \infty$ in the region $(-\pi-2\delta^s) < \sigma < (\pi-2\delta^s)$. Since $\underline{A}$ has a zero eigenvalue,

Eq. (4.37) can be reduced to the simplest particular case corresponding to (2.45) by defining $\tilde{x} = x_2$, $\xi = (D/M)x_1 + x_2$. Hence

$$\begin{aligned}\dot{\tilde{x}} &= -(D/M)\tilde{x} - \frac{1}{M} f(\sigma) \\ &= -\frac{D}{M}\tilde{x} - \phi(\sigma) \\ \dot{\xi} &= -\phi(\sigma) \\ \sigma &= -\frac{M}{D}\tilde{x} + \frac{M}{D}\xi.\end{aligned} \tag{4.39}$$

Note that $\phi(\sigma)$ has been defined as $\frac{1}{M} f(\sigma)$. Equation (4.39) is put in the matrix form as

$$\begin{aligned}\begin{bmatrix} \dot{\underline{x}} \\ \dot{\xi} \end{bmatrix} &= \begin{bmatrix} \underline{A} & 0 \\ \underline{0} & 0 \end{bmatrix} \begin{bmatrix} \underline{x} \\ \xi \end{bmatrix} + \begin{bmatrix} \underline{b} \\ 1 \end{bmatrix} u \\ u &= -\phi(\sigma) \\ \sigma &= \underline{c}^T \underline{x} + d\xi\end{aligned} \tag{4.40}$$

to conform to the notation in Eq. (2.45) where $\underline{x} = \tilde{x}$, $\underline{A} = -D/M$, $\underline{b} = 1$, $d = \frac{M}{D}$, $\underline{c} = -\frac{M}{D}$. Applying Popov's inequality given by (2.49) with $k = \infty$, we have

$$\text{Re}\left[1 + j\omega q \frac{1}{j\omega(j\omega + \frac{D}{M})} \right] \geq 0 \tag{4.41}$$

i.e. $$\frac{q\frac{D}{M} - 1}{\frac{D^2}{M^2} + \omega^2} \geq 0 \quad \text{for all } \omega \geq 0$$

This is satisfied for all $q\frac{D}{M} \geq 1$, i.e.

$$q = \frac{M}{D} n \quad \text{where} \quad n \geq 1.$$

4.6.1 Construction of Lyapunov Function

Kalman's construction procedure explained in Sec. 2.15.4 is now applied. For this system $\psi(s) = s + \frac{D}{M}$

Step 1: From Popov's criterion $q = \frac{Mn}{D}$ where $n \geq 1$.

Step 2: $W(\omega) = \dfrac{(q\frac{D}{M} - 1)}{(\frac{D^2}{M^2} + \omega^2)} (j\omega + \frac{D}{M})(-j\omega + \frac{D}{M}) = n - 1.$

Step 3: Factorization of $W(\omega) = \theta(j\omega)\theta(-j\omega)$ yields trivially $\theta(j\omega) = \sqrt{n-1}$.

Step 4: $r = q(d+\underline{c}^T\underline{b}) = 0.$

Step 5: $\nu(z) = -\theta(z) = -\sqrt{n-1}.$

Step 6: $u = -\sqrt{n-1}.$

Step 7: The Lyapunov matrix equation corresponding to $\underline{A}^T\underline{P} + \underline{P}\underline{A} = -\underline{u}\underline{u}^T$ becomes $-\frac{D}{M}p_{11} + p_{11}(-\frac{D}{M}) = -(n-1)$ where $\underline{P} = p_{11}$. Solving for p_{11} yields $p_{11} = \frac{n-1}{2}\frac{M}{D}$.

From (2.55) we get,

$$V(\tilde{x},\xi) = \underline{X}^T\underline{P}\underline{X} + \frac{1}{2} d\xi^2 + q \int_0^\sigma \phi(\sigma)d\sigma$$

$$= \frac{n-1}{2}\frac{M}{D}\tilde{x}^2 + \frac{1}{2}\frac{M}{D}\xi^2 + \frac{nM}{D}\int_0^\sigma \frac{1}{M} f(\sigma)d\sigma \tag{4.42}$$

Factoring out the constant factor M/D,

$$V(\tilde{x},\xi) = \frac{n-1}{2} x^2 + \frac{1}{2}\xi^2 + \frac{n}{M}\int_0^\sigma f(\sigma)d\sigma$$

$$= \frac{1}{2}\frac{D^2}{M^2} x_1^2 + \frac{n}{2} x_2^2 + \frac{D}{M} x_1 x_2 + \frac{n}{M}\int_0^\sigma f(\sigma)d\sigma$$

Multiplying by M/n and expressing in terms of x_1 and x_2, we get

$$V(x_1,x_2) = \frac{1}{2} M x_2^2 + \frac{1}{2}\frac{D^2}{Mn} x_1^2 + \frac{D}{n} x_1 x_2 + \int_0^\sigma f(\sigma)d\sigma \tag{4.43}$$

This is a fairly general expression for the Lyapunov function. Different choices of n yield different Lyapunov functions. $n = \frac{D}{\beta}$ with $0 \leq \beta \leq D$ gives Mansour's Lyapunov function [4]. The choice of $n = 1$ leads to

$$V(x_1,x_2) = \frac{1}{2} M(\frac{D}{M} x_1 + x_2)^2 + \int_0^{\sigma} f(\sigma)d\sigma \qquad (4.45)$$

which is equivalent to (4.22) obtained by the variable gradient method.

To compute the region of attraction, we use the method due to Walker and McLamroch [8].

Consider the system (2.43)

$$\dot{\underline{x}} = \underline{\underline{A}}\underline{x} - \underline{b}\phi(\sigma)$$

$$\sigma = \underline{c}^T\underline{x}$$

whose V(x) is given by (2.54) as

$$V(\underline{x}) = \underline{x}^T\underline{\underline{P}}\underline{x} + q \int_0^{\sigma} \phi(\sigma)d\sigma$$

Since the nonlinearity $\phi(\sigma)$ violates the sector condition at ℓ_1,ℓ_2, define

$$M_i = \text{Min } V(\underline{x}) \qquad (i=1,2) \qquad (4.45)$$

subject to $\underline{c}^T\underline{x} = \ell_i$

Note that this minimum takes place at a point $\underline{x}_0$ at which $\underline{c}^T\underline{x}_0 = \ell_i$ and $\underline{\nabla V}(\underline{x}_0)$ is orthogonal to the hyperplane $\underline{c}^T\underline{x} = \ell_i$ (i=1,2). Using Lagrange multipliers, we can evaluate $\underline{x}_0 = \ell_i \underline{P}^{-1}\underline{c}/\underline{c}^T\underline{P}^{-1}\underline{c}$ and consequently $M_i = \ell_i^{\,2}/\underline{c}^T\underline{P}^{-1}\underline{c} + q \int_0^{\sigma} \phi(\sigma)d\sigma$. The region of attraction is defined as

$$V(\underline{x}) < C \quad \text{where} \quad C = \text{Min}(M_1,M_2)$$

In the power system problem, $\ell_1 = \pi - 2\delta^s$, $\ell_2 = -\pi - 2\delta^s$ and hence $|\ell_1| < |\ell_2|$. Therefore, by inspection, the region of attraction is

$$V(x_1,x_2) < M_1$$

where

$$M_1 = \frac{(\pi-2\delta^s)^2}{\underline{c}^T\underline{P}^{-1}\underline{c}} + q \int_0^{\pi-2\delta^s} \phi(\sigma)d\sigma \qquad (4.46)$$

We now apply this result to (4.37) whose Lyapunov function $V(\underline{x})$ is given by (4.43). $\underline{P}$ is obtained from (4.43) as

$$\underline{P} = \begin{bmatrix} \frac{1}{2}\frac{D^2}{Mn} & \frac{D}{2n} \\ \frac{D}{2n} & \frac{M}{2} \end{bmatrix} \qquad \underline{P}^{-1} = \frac{4}{(n-1)}\frac{n^2}{D^2}\begin{bmatrix} \frac{M}{2} & -\frac{D}{2n} \\ -\frac{D}{2n} & \frac{D^2}{2Mn} \end{bmatrix}$$

and
$$\frac{\ell_1^2}{\underline{c}^T\underline{P}^{-1}\underline{c}} = \frac{(\pi-2\ {}^{s})^2}{2M}\,\frac{D^2(n-1)}{n^2}\ .$$

The integral term in (4.46) is identified from (4.43) and (4.38) as

$$\int_0^{\pi-2\delta^s} P_e(\sin(\sigma+\delta^s)-\sin\delta^s)\,d\sigma = [2P_e\cos\delta^s - P(\pi-2\delta^s)]$$

The region of attraction is given by

$$\frac{1}{2}\frac{D^2}{Mn}x_1^2 + \frac{M}{2}x_2^2 + \frac{D}{n}x_1x_2 + \int_0^{\sigma} f(\sigma)\,d\sigma$$
$$< \frac{(\pi-2\delta^s)^2}{2M}D^2\left(\frac{(n-1)}{n^2}\right)$$
$$+ [2P_e\cos\delta^s - P(\pi-2\delta^s)] \tag{4.47}$$

For $n = 1$, $D = 0$, this becomes

$$\frac{1}{2}M x_2^2 + \int_0^{\sigma} f(\sigma)\,d\sigma < (2P_e\cos\delta^s - P(\pi-2\delta^s)) \tag{4.48}$$

It may be verified from (4.13) that since $\delta^u = \pi - \delta^s$,

$$V(\delta^u,0) = -P(\pi-2\delta^s) + 2P_e\cos\delta^s \tag{4.49}$$

Hence, inequality (4.48) is equivalent to the equal area criterion and the region of stability obtained by the method of Walker and McLamroch [8] is equivalent to obtaining C by evaluating $V(\underline{x})$ at the nearest unstable equilibrium point. This aspect was utilized in much of the work since 1965 in obtaining region of attraction for multimachine power systems.

The methodology using Kalman's procedure as well as constructing the region of attraction has been extended to higher order systems such as simple governor representation as in Ref. [7].

4.7 Lyapunov Function for Multimachine Power Systems

The various types of models for multimachine power systems have been discussed in Chapter III. Now we shall discuss how to derive the Lyapunov functions for these models. Most of the methods discussed for the single machine cases have been extended to multimachine cases but from the point of view of application, two types of Lyapunov functions which are general for the multimachine case seem to be of interest. These are (i) energy type Lyapunov functions which are analogous to those discussed in Sec. 4.3 but including transfer conductances but no mechanical damping, (ii) Lyapunov functions based on the multivariable Popov criterion.

4.7.1 Energy Type Lyapunov Function [10]

Although this function has been derived in many different ways since originally proposed by Aylett [9], we will adopt the COA formulation discussed in Sec. 3.5 of Chapter III. The COA formulation as opposed to the machine angle formulation has the advantage that the various terms in the Lyapunov function can be given physical meaning analogous to the single machine case. Neglecting damping, the equations are given by (3.28) and (3.29) with $D_i \equiv 0$ $(i=1,2,\ldots,n)$

$$M_T\ddot{\delta}_0 = P_{COA} \tag{4.50}$$

$$M_i\ddot{\theta}_i = P_i - P_{ei} - \frac{M_i}{M_T} P_{COA} \tag{4.51}$$

where $P_{COA} = \sum_{i=1}^{n} P_i - 2\sum_{i=1}^{n}\sum_{j=i+1}^{n} D_{ij}\cos\delta_{ij}$

$$\delta_0 = \frac{1}{M}\sum_{i=1}^{n} M_i\delta_i \quad \text{and} \quad M_T = \sum_{i=1}^{n} M_i$$

Multiply the ith swing equation in (4.51) by $\dot{\theta}_i$ and form the sum

$$\sum_{i=1}^{n} [M_i\dot{\tilde{\omega}}_i - P_i + P_{ei} + \frac{M_i}{M_T} P_{COA}]\dot{\theta}_i \qquad (4.52)$$

Using the inequalities $C_{ij} = C_{ji}$ and $D_{ij} = D_{ji}$ we obtain

$$\sum_{i=1}^{n} \sum_{\substack{j=1 \\ \neq i}}^{n} C_{ij}\sin\theta_{ij}\ \dot{\theta}_i = \sum_{i=1}^{n-1} \sum_{j=i+1}^{n} C_{ij}\sin\theta_{ij}\ \dot{\theta}_{ij} \qquad (4.53)$$

$$\sum_{i=1}^{n} \sum_{\substack{j=1 \\ \neq i}}^{n} D_{ij}\cos\theta_{ij}\ \dot{\theta}_i = \sum_{i=1}^{n-1} \sum_{j=i+1}^{n} D_{ij}\cos\theta_{ij}(\dot{\theta}_i+\dot{\theta}_j) \qquad (4.54)$$

Integrate (4.52) with respect to time using as a lower limit $t = t_s$ where $\underline{\tilde{\omega}}(t_s) = 0$ and $\underline{\theta}(t_s) = \underline{\theta}^s$ is the stable equilibrium point. This yields

$$V = \frac{1}{2} \sum_{i=1}^{n} M_i\tilde{\omega}_i^2 - P_i(\theta_i-\theta_i^s)$$

$$- \sum_{i=1}^{n-1} \sum_{j=i+1}^{n} [C_{ij}(\cos\theta_{ij}-\cos\theta_{ij}^s)$$

$$- \int_{\theta_i^s+\theta_j^s}^{\theta_i+\theta_j} D_{ij}\cos\theta_{ij}\ d(\theta_i+\theta_j) \qquad (4.55)$$

This procedure of arriving at the V-function is the n-dimensional analog of deriving a first integral of motion for the swing equation discussed in Sec. 4.3. Note that in (4.52) the contribution of the last term is zero since

$$\sum_{i=1}^{n} \frac{M_i}{M_T} P_{COA}\ \dot{\theta}_i = P_{COA} \sum_{i=1}^{n} \frac{M_i\tilde{\omega}_i}{M_T} = 0$$

In the V function given by (4.55), the last term is seen to be path dependent. Since $D_{ij} = E_iE_jG_{ij}$, where G_{ij} is the transfer conductance between internal buses i and j, it is the generally held view that transfer conductances prevent the construction of an analytical Lyapunov function. Hence (4.55) as it stands is of limited value. However, during the faulted state since we are integrating the faulted equations, we know $\underline{\theta}$ and hence the integral term in (4.55) can be computed using the trapezoidal rule. The limitation of the V-function given by (4.55) arises from the need to compute C which defines the region of attraction for the post-fault system. In subsequent work, we shall refer to C as V_{cr} to conform to the prevailing terminology. One of the methods of obtaining V_{cr} is to compute the V-function at the unstable equilibrium point corresponding to the mode of instability, i.e. $V_{cr} = V(\underline{\theta}^u, 0)$. Hence, the upper limit of the integral in Eq. (4.55) is $\underline{\theta}^u$. Since the trajectory from $\underline{\theta}^s$ to $\underline{\theta}^u$ is not known, an approximation in the form of a linear path angle from $\underline{\theta}^s$ to $\underline{\theta}^u$ is made. (This aspect will be discussed fully in Chapter V.) Hence, an approximate approximation for V_{cr} can be obtained. Because of the non-integrability of one of the terms in V, it is not possible to judge analytically the sign definiteness of V and $\dot{V}$. From physical considerations such as the existence of the post-fault SEP, one can argue that V > 0, but the same physical arguments cannot be extended to say anything about $\dot{V}$. However, as will be pointed out in Chapter V, this particular form of V-function, also called the energy integral or simply the energy function, has been successfully used in actual large scale power system examples [10] and in the post-fault state except for the small amount of dissipation V = constant, i.e. $\dot{V} = 0$.

Energy Function Neglecting Transfer Conductances

If transfer conductances are neglected in the model (4.50) and (4.51), then $G_{ij} = 0$ and $\sum_{i=1}^{n} P_i = 0$. The V-function (4.55) then becomes

$$V = \frac{1}{2}\sum_{i=1}^{n} M\tilde{\omega}_i^2 - \sum_{i=1}^{n} P_i(\theta_i - \theta_i^s) - \sum_{i=1}^{n-1}\sum_{j=i+1}^{n} [C_{ij}(\cos\theta_{ij} - \cos\theta_{ij}^s)] \tag{4.56}$$

It can be shown that this V-function is equivalent to

$$V = \sum_{i=1}^{n-1} \sum_{j=i+1}^{n} [\frac{1}{2M_T} M_i M_j (\omega_i - \omega_j)^2$$

$$- \frac{1}{M_T} (P_i M_j - P_j M_i)(\delta_{ij} - \delta_{ij}^s) - C_{ij} \cos(\delta_{ij} - \delta_{ij}^s)] \qquad (4.57)$$

This V-function has been used by a large number of research workers. Aylett called it the energy integral [9], and later on it has been derived by DiCaprio and Saccomano [30] and Ribbens-Pavella [13,26]. It was also shown that V is indeed a Lyapunov function satisfying $V > 0$ and $\dot{V} = 0$. Hence, the post-fault SEP $\underline{\theta}^s$ is Lyapunov stable. In the presence of mechanical damping, we can indeed show that $\dot{V} \leq 0$ and asymptotic stability in a region around $\underline{\theta}^s$ is ensured since $\dot{V} \neq 0$ on a trajectory inside that region. One can verify $V > 0$ in a number of ways. From physical considerations since V represents the energy integral, it has to be greater than zero. Mathematically speaking, it can be shown that V is equal to the sum of a quadratic term and the sum of integrals of nonlinearities, each of which lies in the first and third quadrants in a region around the origin. Hence, $V > 0$.

4.7.2 Lyapunov Function Using Multivariable Popov Criterion (Zero Transfer Conductance)

A more systematic way of developing Lyapunov functions by the multivariable Popov criterion is presented in this section. As a special case, it includes the energy type Lyapunov function of the previous section. The Luré form of the mathematical model discussed in Sec. 3.6 of Chapter III is the most convenient to use. In the literature, Lyapunov functions have been developed analytically only for systems with zero transfer conductance. With non-negligible transfer conductances, there are results which take them into account only partially. We therefore consider first the system with zero transfer conductance with non-uniform damping.

System with Zero Transfer Conductances (non-uniform damping)

The state model is given by Eq. (3.40) as

$$\dot{\underline{x}} = \underline{A}\underline{x} - \underline{B}\underline{f}(\underline{\sigma})$$
$$\underline{\sigma} = \underline{C}\underline{x} \qquad (4.58)$$

where
$$\underline{A} = \begin{array}{c} \begin{array}{cc} n & n-1 \end{array} \\ \left[\begin{array}{c|c} -\underline{M}^{-1}\underline{D} & \underline{0} \\ \hline \underline{K}_1^T & \underline{0} \end{array}\right] \end{array} \qquad \underline{B} = \begin{array}{c} m \\ \left[\begin{array}{c} \underline{M}^{-1}\underline{K} \\ \hline \underline{0} \end{array}\right] \end{array}$$

$$\underline{C} = \begin{array}{c} \begin{array}{cc} n & n-1 \end{array} \\ \left[\begin{array}{c|c} \underline{0} & \underline{S} \end{array}\right] \end{array} m = \begin{array}{c} \begin{array}{cc} n & n-1 \end{array} \\ \left[\begin{array}{c|c} \underline{0} & \underline{I} \\ \hline \underline{0} & \underline{J}^T \end{array}\right] \end{array} \begin{array}{l} n-1 \\ m-n+1 \end{array}$$

Furthermore, $\underline{K}_1\underline{S}^T = \underline{K}$.

The nonlinearity is given by (3.44) as

$$f_k(\underline{\sigma}) = f_k(\sigma_k) = E_pE_qB_{pq}[\sin(\sigma_k+\delta_{pq}^s) - \sin\delta_{pq}^s] \qquad (4.59)$$

For $k=1,2,\ldots,n-1$, $\sigma_k = \delta_{pn} - \delta_{pn}^s$ $(p=1,2,\ldots,n-1)$

For $k=(n-1),\ldots,m$, $\sigma_k = \delta_{pq} - \delta_{pq}^s$ $(p=1,2,\ldots,n-2$, $q=2,3,\ldots,n-1$, $q>p)$

The nonlinearity $f_k(\sigma_k)$ satisfies the sector condition $f_k(\sigma_k)\sigma_k > 0$, $k=1,2,\ldots,m$, in an interval around σ_k which we shall denote by

$$\ell_{k2} < \sigma_k < \ell_{k1}, \quad \ell_{k1} > 0, \quad \ell_{k2} < 0$$

The transfer function matrix $\underline{W}(s)$ is given by $\underline{W}(s) = \underline{C}(s\underline{I}-\underline{A})^{-1}\underline{B}$ (Fig. 3.4). To apply the multivariable Popov criterion of Sec. 2.16, more popularly known as the Moore-Anderson Theorem, we must find $\underline{N}$ and $\underline{Q}$ such that $\underline{Z}(s) = (\underline{N}+\underline{Q}s)\underline{W}(s)$ is positive real. We shall take both $\underline{N}$ and $\underline{Q}$ to be diagonal of the form $\underline{N} = n\underline{I}$ and $\underline{Q} = q\underline{I}$ so that

$$\underline{Z}(s) = (n+qs)\underline{W}(s) \qquad (4.60)$$

The conditions for $\underline{Z}(s)$ to be positive real are [15]:

(i) $\underline{Z}(s)$ has elements which are analytic for $Re(s) > 0$.

(ii) $\underline{Z}^*(s) = \underline{Z}(s^*)$. (4.61)

(iii) $\underline{Z}^T(s^*) + \underline{Z}(s)$ is positive semi-definite for $Re(s) > 0$.

The first two conditions clearly hold. We need to find n and q such that (iii) holds. From (3.39)

$$\underline{W}(s) = \frac{1}{s}\,[\underline{K}^T(s\underline{I}+\underline{\lambda})^{-1}\underline{M}^{-1}\underline{K}] \tag{4.62}$$

where $\underline{\lambda} = \underline{M}^{-1}\underline{D}$. $\underline{W}(s)$ can be rewritten as

$$\underline{W}(s) = \frac{1}{s}\,\underline{K}^T(s\underline{M}+\underline{D})^{-1}\underline{K} \tag{4.63}$$

where $(s\underline{M}+\underline{D})^{-1}$ is a diagonal matrix. Therefore, $\underline{W}(s)$ is symmetric and hence $\underline{Z}(s)$ is also symmetric. Thus, condition (iii) of (4.61) becomes

$$\begin{aligned}\underline{Z}^T(s^*) + \underline{Z}(s) &= \underline{Z}(s^*) + \underline{Z}(s)\\ &= 2Re[\underline{Z}(s)] \\ &= 2\underline{K}^T[Re\{\frac{n + qs}{s(sM_i+D_i)}\}]\underline{K} \geq 0 \text{ for } Re(s) > 0\end{aligned} \tag{4.64}$$

The above condition will be satisfied if n and q are chosen such that the diagonal matrix

$$Re[\frac{n + qs}{s(sM_i+D_i)}] \geq 0 \text{ for } Re(s) > 0 \tag{4.65}$$

This clearly requires each term to be non-negative and a sufficient condition on n and q is

$$q \geq n\,\frac{M_i}{D_i} \tag{4.66}$$

Equation (4.66) also follows from the fact that $\underline{K}$ is the incidence matrix of an n-node passive network and $\underline{W}(s)$ is its admittance matrix [19]. There is another condition to be satisfied in the Moore-Anderson Theorem, namely that the choice of $\underline{N}$ and $\underline{Q}$ does not cause pole-zero cancellation in $\underline{W}(s)$. This requires that $\frac{q}{n} \neq \infty$ or $= \frac{M_i}{D_i}$. However, as we shall see later this condition is relaxed in many cases. Hence, a valid range of n and q is $q \geq n \frac{M_i}{D_i}$ for all i and both q and n are finite. Note the similarity of this condition to that derived in Sec. 4.6 for the single machine case, namely $q \geq \frac{M}{D} n$ with $n \geq 1$.

With $\underline{N}$ and $\underline{Q}$ chosen, there exists by the Moore-Anderson Theorem [15] a Lyapunov function of the form

$$V(\underline{x}) = \underline{x}^T \underline{P}\underline{x} + 2 \int_0^{\underline{C}\underline{x}} \underline{f}^T(\underline{\sigma})\underline{Q}\, d\underline{\sigma} \tag{4.67}$$

where $\underline{P}$ is a positive definite matrix satisfying the equations

$$\underline{A}^T\underline{P} + \underline{P}\underline{A} = -\underline{L}\,\underline{L}^T \tag{4.68}$$

$$\underline{P}\underline{B} = \underline{C}^T\underline{N} + \underline{A}^T\underline{C}^T\underline{Q} - \underline{L}\,\underline{W}_o \tag{4.69}$$

$$\underline{W}_o^T\underline{W} = \underline{Q}\underline{C}\underline{B} + \underline{B}^T\underline{C}^T\underline{Q} \tag{4.70}$$

where $\underline{L}$ and $\underline{W}_o$ are auxiliary matrices. In the power system model, it can be verified that $\underline{C}\underline{B} = \underline{0}$. Hence, $\underline{W}_o$ can be chosen to be zero. Equations (4.68) - (4.70) become with $\underline{N} = n\underline{I}$ and $\underline{Q} = q\underline{I}$

$$\underline{A}^T\underline{P} + \underline{P}\underline{A} = -\underline{L}\,\underline{L}^T \tag{4.71}$$

$$\underline{P}\underline{B} = n\underline{C}^T + q\underline{A}^T\underline{C}^T \tag{4.72}$$

Let

$$\underline{P} = \begin{array}{c} \\ n \\ n-1 \end{array}\!\!\begin{array}{c} \begin{array}{cc} n & n-1 \end{array} \\ \begin{bmatrix} \underline{P}_1 & \underline{P}_2^T \\ \underline{P}_2 & \underline{P}_3 \end{bmatrix} \end{array} \qquad \underline{L} = \begin{bmatrix} \underline{L}_1 \\ \underline{L}_2 \end{bmatrix}\begin{array}{c} n \\ n-1 \end{array}$$

Equations (4.71) and (4.72) then reduce to

$$\left[\begin{array}{c|c} -\underline{P}_1\underline{M}^{-1}\underline{D} - \underline{D}\underline{M}^{-1}\underline{P}_1 + \underline{P}_2^T\underline{K}_1^T + \underline{K}_1\underline{P}_2 & -\underline{D}\underline{M}^{-1}\underline{P}_2^T + \underline{K}_1\underline{P}_3^T \\ \hline -\underline{P}_2\underline{M}^{-1}\underline{D} + \underline{P}_3\underline{K}_1^T & \underline{0} \end{array}\right] = -\left[\begin{array}{c|c} \underline{L}_1\underline{L}_1^T & \underline{L}_1\underline{L}_2^T \\ \hline \underline{L}_2\underline{L}_1^T & \underline{L}_2\underline{L}_2^T \end{array}\right] \tag{4.73}$$

$$\underline{P}_1\underline{M}^{-1}\underline{K} = q\underline{K} \tag{4.74}$$

$$\underline{P}_2\underline{M}^{-1}\underline{K} = n\underline{S}^T \tag{4.75}$$

where $\underline{K} = \underline{K}_1\underline{S}^T$

$\underline{L}_2$ is chosen as zero matrix from (4.73). Hence, we have from (4.73)

$$-\underline{P}_1\underline{M}^{-1}\underline{D} - \underline{D}\underline{M}^{-1}\underline{P}_1 + \underline{P}_2^T\underline{K}_1^T + \underline{K}_1\underline{P}_2 = -\underline{L}_1\underline{L}_1^T \leq 0 \tag{4.76}$$

$$-\underline{P}_2\underline{M}^{-1}\underline{D} + \underline{P}_3\underline{K}_1^T = \underline{0} \tag{4.77}$$

Euqations (4.74) to (4.77) are now solved for $\underline{P}_1$, $\underline{P}_2$, $\underline{P}_3$ as follows: From (4.77) we get

$$\underline{P}_2 = \underline{P}_3\underline{K}_1^T\underline{D}^{-1}\underline{M} \tag{4.78}$$

From (4.74) we get

$$(\underline{M}^{-1}\underline{P}_1\underline{M}^{-1} - q\underline{M}^{-1})\underline{K} = 0 \tag{4.79}$$

$\underline{K}$ has the structure of an incidence matrix and has only +1 and -1 in each of its columns. Therefore

$$\underline{M}^{-1}\underline{P}_1\underline{M}^{-1} - q\underline{M}^{-1} = r\underline{U} \tag{4.80}$$

where $\underline{U}$ is a unit matrix with all entries equal to unity and r is a constant. This yields

$$\underline{P}_1 = q\underline{M} + r\underline{MUM} \tag{4.81}$$

Now eliminate $\underline{P}_2$ from (4.75) and (4.78) to get

$$\underline{K}_1\underline{P}_3\underline{K}_1^T\underline{D}^{-1}\underline{K} = n\underline{K} \tag{4.82}$$

Hence, utilizing the structure of $\underline{K}$ again we get

$$\underline{K}_1\underline{P}_3\underline{K}_1^T = n\underline{D} + s\underline{DUD} \tag{4.83}$$

where s is a constant. It may be verified that Eq. (4.83) has the solution

$$\underline{P}_3 = n\underline{D}_{n-1} + s\underline{D}_{n-1}\underline{U}\underline{D}_{n-1} \tag{4.84}$$

where $\underline{D}_{n-1}$ is the matrix $\underline{D}$ with n^{th} row and n^{th} column deleted. Substituting $\underline{P}_1$ and $\underline{P}_2$ from (4.81) and (4.78) respectively, the matrix inequality (4.76) becomes

$$-2q\underline{D} - r(\underline{MUD}+\underline{DUM}) + \underline{MD}^{-1}\underline{K}_1\underline{P}_3\underline{K}_1^T$$
$$+ \underline{K}_1\underline{P}_3\underline{K}_1^T\underline{D}^{-1}\underline{M} \leq 0 \tag{4.85}$$

since $\underline{P}_3$ is symmetric.

Substituting the expression for $\underline{K}_1\underline{P}_3\underline{K}_1^T$ from (4.83) into (4.85) we get

$$2(-q\underline{D}+n\underline{M}) - (r-s)(\underline{MUD}+\underline{DUM}) \leq 0 \tag{4.86}$$

Dividing by q and letting $\mu = (r-s)/q$, (4.86) becomes

$$2(-\underline{D} + \frac{n}{q}\underline{M}) - \mu(\underline{MUD}+\underline{DUM}) \leq 0$$

i.e. $\underline{F}(\mu) \leq 0$ (4.87)

The determinant of the left-hand side of the inequality (4.87) is a quadratic in μ [1,20]. The matrix $\underline{A}^T\underline{P} + \underline{PA}$ is negative semi-definite if and only if the matrix $\underline{F}(\mu)$ is negative semi-definite. Let

$$\det(\underline{F}(\mu)) = a_2\mu^2 + a_1\mu + a_0 \tag{4.88}$$

$$a_2 = -2^n \prod_{i=1}^{n} (-D_i + n\frac{M_i}{q})$$

$$\left[\sum_{i=1}^{n-1}\sum_{j=i+1}^{n} \frac{1}{4}\left[\frac{(D_iM_j-D_jM_i)^2}{(-D_i + n\frac{M_i}{q})(-D_j + n\frac{M_j}{q})}\right]\right]$$

$$A_1 = 2^n \sum_{i=1}^{n} (-D_i + n\frac{M_i}{q}) \sum_{i=1}^{n} \frac{(-D_iM_i)}{(-D_i + n\frac{M_i}{q})}$$

$$a_0 = 2^n \sum_{i=1}^{n} (-D_i + n\frac{M_i}{q}) \tag{4.89}$$

Let μ_1 be the largest negative value of satisfying $\det(\underline{F}(\mu)) = 0$. Then Eq. (4.86) is satisfied if $\mu_1 \leq \mu \leq 0$ since $\underline{F}(0) = 2(-\underline{D} + \frac{n}{q}\underline{M})$ which is negative definite. The quadratic equation $\det(\underline{F}(\mu)) = 0$ can be written in a simplified form as

$$\mu^2 \left[\sum_{i-1}^{n-1} \sum_{j=i+1}^{n} \frac{1}{4} \frac{(D_i M_j - D_j M_i)}{(D_i - n\frac{M_i}{q})(D_j - n\frac{M_i}{q})} \right]$$

$$- \mu \sum_{i=1}^{n} \left(\frac{D_i M_i}{D_i - n\frac{M_i}{q}} \right) - 1 = 0 \tag{4.90}$$

Different choices of n and q satisfying the positive real property and different choices of μ lying between 0 and μ_1 will result in various Lyapunov functions. The matrices $\underline{P}_1$, $\underline{P}_2$, $\underline{P}_3$ are given by (4.81), (4.78) and (4.84) respectively.

(i) Lyapunov Function of El-Abiad and Nagappan [11], Gless [16]

Choose $n = 0$, $q = 1$, $r = 0$, $s = 0$. With this choice, $\mu = 0$, $\underline{P}_1 = \underline{M}$, $\underline{P}_2 = \underline{P}_3 = 0$. The Lyapunov function is

$$V(\underline{x}) = \underline{x}^T \left[\begin{array}{c|c} \underline{M} & \underline{0} \\ \hline \underline{0} & \underline{0} \end{array} \right] \underline{x} + 2 \int_0^{\underline{\sigma}} \underline{f}^T(\underline{\sigma})\, d\underline{\sigma} \tag{4.91}$$

Dividing the right-hand side by 2 and reverting to the physical variables

$$V(\underline{x}) = \frac{1}{2} M_i \omega_i^2 + \sum_{i=1}^{m} \int_0^{\sigma_i} f_i(u_i)\, du_i \tag{4.92}$$

(ii) Lyapunov Function of Willems [12]

Choose $n = 0$, $q = 1$, $s = 0$, $r = \mu_1$ where μ_1 is the solution (negative) of the quadratic equation (4.89). This results in the Lyapunov function

$$V(\underline{x}) = \underline{x}^T \left[\begin{array}{c|c} \underline{M} + \mu_1 \underline{MUM} & \underline{0} \\ \hline \underline{0} & \underline{0} \end{array}\right] \underline{x} + 2 \int_0^{\underline{\sigma}} \underline{f}^T(\underline{u})d\underline{u} \qquad (4.93)$$

After dividing by 2 and expressing in the physical variables,

$$V(\underline{x}) = \frac{1}{2} \sum_{i=1}^{n} M_i\omega_i^2 + \frac{\mu_1}{2} \underline{\omega MUM\omega} + \sum_{i=1}^{m} \int_0^{\sigma_i} f_i(u_i)du_i \qquad (4.94)$$

(iii) Lyapunov function of Pai and Murthy [16]

$$n = 1,\ q = \sum_{i=1}^{n} \frac{M_i}{D_i} \qquad r = 0,\ s = -\frac{1}{\Sigma D_i} = -\frac{1}{D_T}$$

$$\underline{P}_1 = \left[\sum_{i=1}^{n} \frac{M_i}{D_i}\right] \underline{M}$$

From (4.84), $\underline{P}_3 = \underline{D}_{n-1} - \frac{1}{D_T} \underline{D}_{n-1}\underline{UD}_{n-1}$ where $\underline{D}_{n-1}$ is $(n-1)\times(n-1)$ diagonal matrix with elements D_i $(i=1,2,\ldots,n-1)$

$$\underline{P}_2 = \underline{P}_3\underline{K}_1^T\underline{D}^{-1}\underline{M}$$

After simplification and dividing by 2, the Lyapunov function reduces to

$$V(\underline{x}) = \frac{1}{2} \sum_{i=1}^{n} \sum_{\substack{j=1 \\ \neq i}}^{n} \left(\frac{M_j}{D_j}\right) M_i\omega_i^2 + \frac{1}{2D_T} \sum_{i=1}^{n} M_i^2\omega_i^2$$

$$+ \frac{1}{2D_T} \sum_{i=1}^{n-1} \sum_{j=i+1}^{n}$$

$$D_iD_j\left(\delta_i - \delta_i^s + \delta_j^s - \delta_j + \frac{M_i}{D_i}\omega_i - \frac{M_j}{D_j}\omega_j\right)^2$$

$$+ \left(\sum_{i=1}^{n} \frac{M_i}{D_i}\right) \sum_{i=1}^{m} \int_0^{\sigma_i} f_i(u_i)du_i \qquad (4.95)$$

(iv) Lyapunov function due to Mansour [4]

Choose $n = 0$, $q = \frac{1}{2}$, $r = s \geq -\frac{1}{D_T}$. It can be shown that

$$V(\underline{x}) = \frac{1}{2}\sum_{i=1}^{n} M_i\omega_i^2 + \frac{r}{2}\left[\sum_{i=1}^{n}(M_i\omega_i + D_i(\delta_i-\delta_i^s))\right]^2 + \sum_{i=1}^{m}\int_0^{\sigma_i} f_i(u_i)du_i \tag{4.96}$$

(v) General Lyapunov function of Willems [19]

Willems [19] has shown that the most general $V(\underline{x})$ is obtained by choosing

$$q \geq n \operatorname*{Max}_i \left(\frac{M_i}{D_i}\right), \qquad r = s \geq -\frac{1}{D_T}\,n$$

$$V(\underline{x}) = \frac{1}{2}\sum_{i=1}^{n}\left(q - n\frac{M_i}{D_i}\right)M_i\omega_i^2$$

$$+ \frac{1}{2}\left(s + \frac{n}{D_T}\right)\left[\sum_{i=1}^{n}(M_i\omega_i + D_i(\delta_i-\delta_i^s))\right]^2$$

$$+ \frac{1}{2}\sum_{i=1}^{n} D_i \sum_{i=1}^{n}$$

$$D_iD_j\left(\delta_i - \delta_i^s + \delta_j^s - \delta_j + \frac{M_i}{D_i}\omega_i - \frac{M_j}{D_j}\omega_j\right)^2$$

$$+ \sum_{i=1}^{m}\int_0^{\sigma_i} f_i(u_i)du_i \tag{4.97}$$

Uniform Damping

Let $\frac{D_i}{M_i} = \lambda$ for all i

The state space model is obtained from Eq. (3.48) by neglecting the transfer conductances, i.e. $\underline{B}_2 \equiv \underline{0}$.

$$\dot{\underline{x}} = \hat{\underline{A}}\underline{x} - \hat{\underline{B}}\ \underline{f}(\underline{\sigma})$$

$$\underline{\sigma} = \hat{\underline{C}}\underline{x} \tag{4.98}$$

$$\underline{x} = [\omega_{1n},\ldots,\omega_{n-1,n} \mid \delta_{1n}-\delta^{s}_{1n},\ldots,\delta_{n-1,n}-\delta^{s}_{n-1,n}]$$

where $\omega_{ij} = \omega_i - \omega_j, \quad \delta_{ij} = \delta_i - \delta_j$

$$\hat{\underline{A}} = \begin{array}{c} \begin{array}{cc} n-1 & n-1 \end{array} \\ \left[\begin{array}{c|c} -\underline{\lambda} & \underline{0} \\ \hline \underline{I} & \underline{0} \end{array}\right] \begin{array}{c} n-1 \\ n-1 \end{array} \end{array} \qquad \hat{\underline{B}} = \left[\begin{array}{c} \underline{K}_1^T \underline{M}^{-1} \underline{K} \\ \hline \underline{0} \end{array}\right] \begin{array}{c} n-1 \\ n-1 \end{array}$$

$$\hat{\underline{C}} = \begin{array}{c} \begin{array}{cc} n-1 & n-1 \end{array} \\ \left[\begin{array}{c|c} \underline{0} & \underline{I} \\ \hline \underline{0} & \underline{J}^T \end{array}\right] \begin{array}{c} n-1 \\ m-n+1 \end{array} \end{array}$$

The matrix equations in the Moore-Anderson Theorem are

$$\begin{aligned} \hat{\underline{A}}^T\underline{P} + \underline{P}\hat{\underline{A}} &= -\underline{L}\ \underline{L}^T \\ \underline{P}\hat{\underline{B}} &= n\hat{\underline{C}}^T + q\ \hat{\underline{A}}^T\ \hat{\underline{C}}^T \end{aligned} \tag{4.99}$$

Let

$$\underline{P} = \begin{array}{c} \begin{array}{cc} n-1 & n-1 \end{array} \\ \begin{array}{c} n-1 \\ n-1 \end{array} \left[\begin{array}{cc} \underline{P}_1 & \underline{P}_2^T \\ \underline{P}_2 & \underline{P}_3 \end{array}\right] \end{array} ; \qquad \underline{L} = \left[\begin{array}{c} \underline{L}_1 \\ \underline{L}_2 \end{array}\right] \begin{array}{c} n-1 \\ n-1 \end{array}$$

The equation (4.99) becomes

$$\left[\begin{array}{c|c} -\underline{P}_1\ \underline{\lambda} - \underline{\lambda}\ \underline{P}_1 + \underline{P}_2^T + \underline{P}_2 & -\underline{\lambda}\ \underline{P}_2^T + \underline{P}_3^T \\ \hline -\underline{P}_2\ \underline{\lambda} + \underline{P}_3 & \underline{0} \end{array}\right]$$

$$= -\left[\begin{array}{c|c} \underline{L}_1\ \underline{L}_1^T & \underline{L}_1\ \underline{L}_2^T \\ \hline \underline{L}_2\ \underline{L}_1^T & \underline{L}_2\ \underline{L}_2^T \end{array}\right] \tag{4.100}$$

$$\underline{P}_1\underline{K}_1^T\underline{M}^{-1}\underline{K} = q\underline{S}^T \qquad (4.101)$$

$$\underline{P}_2\underline{K}_1^T\underline{M}^{-1}\underline{K} = n\underline{S}^T \qquad (4.102)$$

As in the non-uniform damping case, we choose $\underline{L}_2 = 0$ in (4.100) which then reduces to

$$-\underline{P}_1\ \underline{\lambda} - \underline{\lambda}\ \underline{P}_1 + \underline{P}_2^T + \underline{P}_2 \leq 0 \qquad (4.103)$$

$$\underline{P}_3 = \underline{P}_2\ \underline{\lambda} \qquad (4.104)$$

Pre-multiplying (4.101) by $\underline{M}^{-1}\underline{K}_1$ and rearranging, we get

$$(\underline{M}^{-1}\underline{K}_1\underline{P}_1\underline{K}_1^T\underline{M}^{-1} - q\underline{M}^{-1})\underline{K} = \underline{0} \qquad (4.105)$$

Therefore

$$\underline{K}_1\underline{P}_1\underline{K}_1^T = q\underline{M} + r\ \underline{MUM} \qquad (4.106)$$

It can be shown that

$$\underline{P}_1 = q\underline{M}_{n-1} + r\ \underline{M}_{n-1}\underline{UM}_{n-1} \qquad (4.107)$$

where $\underline{M}_{n-1}$ is the $\underline{M}$ matrix with n^{th} row and n^{th} column deleted. We can also show that

$$\underline{P}_2 = n\underline{M}_{n-1} + s\underline{M}_{n-1}\underline{UM}_{n-1} \qquad (4.108)$$

$$\underline{P}_3 = \underline{\lambda}\ \underline{P}_2 \qquad (4.109)$$

Particular choices of n, q, r and s yield different Lyapunov functions used in the literature.

(i) Choosing $n = 0$, $q = 1$, $r = -\frac{1}{M_T}$, $s = 0$ yields the Lyapunov function appearing in the works of Aylett [9], Ribbens-Pavella [13,18], Willems [12]. Then

$$\underline{P}_2 = \underline{P}_3 = 0$$

Denoting the relative velocity subvector $\underline{x}_1^T = (\omega_{1n}, \omega_{2n}, \ldots \omega_{n-1,n})$, the Lyapunov function is given by

$$V(\underline{x}) = \underline{x}_1^T[\underline{M}_{n-1} - \frac{1}{M_T}\underline{M}_{n-1}\underline{U}\underline{M}_{n-1}]\underline{x}_1 + 2\int_{\underline{0}}^{\underline{\sigma}}\underline{f}^T(\underline{u})d\underline{u} \quad (4.110)$$

Dividing by 2

$$V(\underline{x}) = \frac{1}{2M_T}\sum_{i=1}^{n}\sum_{j=i+1}^{n} M_iM_j(\omega_i-\omega_j)^2$$

$$+ \sum_{i=1}^{m}\int_0^{\sigma_i} f_i(u_i)du_i$$

(ii) Choice of $n = \lambda$, $q = 1$, $r = s = 0$ yields the Lyapunov function in Ref. [14].

$$V(\underline{x}) = \frac{1}{2}\sum_{i=1}^{n} M_i(\omega_i+\lambda(\delta_i-\delta_i^s))^2$$

$$+ \sum_{i=1}^{m}\int_0^{\sigma_i} f_i(\sigma_i)d\sigma_i \quad (4.111)$$

This Lyapunov function may be considered as the n-machine analog of the Lyapunov function for the single m/c case (Eq. 4.44).

(iii) $n = \lambda$, $q = 1$, $r = -\frac{1}{M_T}$, $s = -1/M_T$.

The Lyapunov function is

$$V(\underline{x}) = \frac{1}{2}\frac{1}{M_T}\sum_{i=1}^{n}\sum_{j=i+1}^{n}\{M_iM_j(\omega_i-\omega_j+\lambda(\delta_i-\delta_j-(\delta_i^s-\delta_i^s))\}^2$$

$$+ \sum_{i=1}^{m}\int_0^{\sigma_i} f_i(u_i)du_i \quad (4.112)$$

(iv) $n = \lambda$, $q = 2$, $r = -\frac{2}{M_T}$, $s = -\frac{\lambda}{D_T}$.

The resulting Lyapunov function will have the quadratic part which is the sum of the quadratic parts in (i) and (iii) and the integral term equal to $\sum_{i=1}^{m}\int_0^{\sigma_i} f_i(u_i)du_i$. This Lyapunov function was proposed in Ref. [16].

(v) The most general Lyapunov function for the uniform damping case will be [19]

$$V(\underline{x}) = \frac{1}{2M_T} \sum_{i=1}^{n} \sum_{j=i+1}^{n}$$

$$\{(1 - \frac{a}{\lambda})M_iM_j(\omega_i-\omega_j)^2 + \frac{a}{\lambda} M_iM_j(\omega_i-\omega_j+\lambda(\delta_{ij}-\delta_{ij}^s))^2$$

$$+ E_iE_jY_{ij}[\cos\delta_{ij}-\cos\delta_{ij}^s-(\delta_{ij}-\delta_{ij}^s)\sin\delta_{ij}^s]\} \quad (4.113)$$

for $0 \leq a \leq \lambda$.

This yields the result (i) for $a = 0$, result (iii) for $a = \lambda$, and result (iv) for $a = \frac{\lambda}{2}$.

Zero Damping

The state model is the same as for uniform damping with $\underline{\lambda} \equiv 0$ and hence the Lyapunov functions are readily obtained from the uniform damping case by letting $\underline{\lambda} \equiv 0$.

4.7.3 Lyapunov Functions with Transfer Conductances

The state space model has been derived in Chapter III (Eq. 3.45) for non-uniform damping case as

$$\dot{\underline{x}} = \underline{A}\underline{x} - \underline{B}_1\, \underline{f}(\underline{\sigma}) - \underline{B}_2\, \underline{g}(\underline{\sigma})$$
$$\underline{\sigma} = \underline{C}\underline{x} \quad (4.114)$$

The components of $\underline{f}(\underline{\sigma})$, i.e. $f_i(\sigma_i)$ satisfy the sector conditions in an interval around the origin whereas the terms $\underline{g}(\sigma)$ due to the transfer conductances do not satisfy such a property. Hence, application of the multivariable Popov criterion is not possible. Various ways to circumvent the difficulty have been reported in the literature [20,21,22]. One way is to define the nonlinearities

$$h_i(\sigma_i) = f_i(\sigma_i) + g_i(\sigma_i)$$

$$\ell_i(\sigma_i) = f_i(\sigma_i) - g_i(\sigma_i)$$

Equations (4.114) can then be cast in the form as

$$\dot{\underline{x}} = \underline{A}\underline{x} - [\underline{D}_1 \quad \underline{D}_2]\begin{bmatrix} \underline{h}(\sigma) \\ \underline{\ell}(\underline{\sigma}) \end{bmatrix} \qquad \underline{\sigma} = \underline{C}\underline{x} \quad (4.115)$$

Equation (4.115) now has a Luré structure where both $h_i(\sigma_i)$ and $\ell_i(\sigma_i)$ satisfy the sector conditions. Gudaru [23] tried to apply the Moore-Anderson Theorem to (4.115) but Willems [24] has shown the result to be in error by proving that it is not possible to find an $\underline{N}$ and $\underline{Q}$ to satisfy the theorem. Hence, there have been a number of efforts at approximating the effects of transfer conductances [25-27].

One of the methods [25] is to add and subtract $\underline{B}_1\ \underline{g}(\underline{\sigma})$ from (4.114) which results in

$$\dot{\underline{x}} = \underline{\underline{A}}\underline{x} - \underline{\underline{B}}_1\ \underline{h}(\underline{\sigma}) - (\underline{\underline{B}}_2 - \underline{\underline{B}}_1)\underline{g}(\underline{\sigma})$$

$$\underline{\sigma} = \underline{\underline{C}}\underline{x} \tag{4.116}$$

If the term $(\underline{B}_2-\underline{B}_1)\underline{g}(\underline{\sigma})$ which destroys the Luré structure can be neglected, the resulting approximate model is amenable to analysis via the Moore-Anderson Theorem. The Lyapunov function will be

$$V(\underline{x}) = \underline{x}^T\underline{\underline{P}}\underline{x} + \int_0^{\sigma} \underline{h}(\underline{u})\underline{Q}d\underline{u} \tag{4.117}$$

$$h_i(\underline{\sigma}) = h_i(\sigma_i)$$

$$= E_pE_qY_{pq}[\sin(\sigma_i+\delta^s_{pq}+\theta_{pq}) - \sin(\delta^s_{pq}+\theta_{pq})]$$

This nonlinearity satisfies the sector condition $0 < \dfrac{f_i(\sigma_i)}{\sigma_i} < \infty$ in the interval $-\pi - 2(\delta^s_{pq}+\theta_{pq}) < \sigma_i < \pi - 2(\delta^s_{pq}+\theta_{pq})$. The other method is to have an over description of the system in an n(n-1) state space and then neglect some G_{ij}'s [26]. Still another method is to use an approximation to the energy type of Lyapunov function [27].

In the method involving energy functions and COA formulation, [10] $V(\underline{x})$ is retained in the form of a path dependent integral (4.55). The integral is computed depending on the faulted trajectory and using the trapezoidal rule. In computing $V(\underline{x}^u)$ a linear path angle is assumed between $\underline{\theta}^s$ and $\underline{\theta}^u$. This method will be discussed in detail in Chapter V.

4.8 Lyapunov Functions for Multimachine Systems Including Flux Decay Effects [28]

So far, in the classical model of multimachine systems, the internal voltages have been assumed to be of constant magnitude. This is equivalent to neglecting the effect of flux decay in the machine. In this section, we will relax this assumption and construct a Lyapunov function within the framework of the Moore-Anderson Theorem. The mathematical model is given in Sec. 3.8 of Chapter III.

The transfer function of the linear part of the system is given by

$$\begin{aligned}\underline{W}(s) &= \underline{C}(S\underline{I}-\underline{A})^{-1}\underline{B} \\ &= \begin{bmatrix} \underline{W}_1(s) & \underline{0} \\ \underline{0} & \underline{W}_2(s) \end{bmatrix} \\ &= \begin{bmatrix} \underline{K}^T[s(s\underline{I}+\underline{\lambda})]^{-1}\underline{M}^{-1}\underline{K} & \underline{0} \\ \underline{0} & \underline{I}(s\underline{I}+\underline{\alpha})^{-1}\underline{\beta} \end{bmatrix}\end{aligned} \tag{4.118}$$

We use a modified version of the Moore-Anderson Theorem due to Desoer and Wu [29] for the stability criteria.

The system to be considered is of the form given by Eq. (3.54) with $\underline{W}(s)$ in Eq. (4.118) as the transfer function matrix.

The nonlinearity $\underline{F}(\underline{\sigma})$ is assumed to satisfy the following conditions:

(i) $\underline{F}(\underline{\sigma})$ is continuous, maps R^m into R^{m+n}.

(ii) For some constant real matrix $\underline{N}$, $\underline{F}(\underline{\sigma})^T\underline{N}\underline{\sigma} \geq 0$ for all $\underline{\sigma} \varepsilon R^m$ and $\underline{F}(\underline{\sigma}) = 0$ if $\underline{\sigma} = 0$.

(iii) There is a scalar function $V_1(\underline{\sigma})$ such that $V_1(\underline{\sigma}) \geq 0$ for all $\underline{\sigma} \varepsilon R^m$, $V_1(\underline{\sigma}) = 0$ for $\underline{\sigma} = 0$, and for some constant real matrix $\underline{Q}$

$$\nabla \underline{V}_1(\underline{\sigma}) = \underline{Q}^T\underline{F}(\underline{\sigma}) \quad \text{for all} \quad \underline{\sigma} \varepsilon R^m$$

If we can find $\underline{N}$ and $\underline{Q}$ to satisfy Moore-Anderson Theorem with the above restrictions, then there exists a Lyapunov function of the form

$$V(\underline{x}) = \underline{x}^T\underline{P}\underline{x} + 2\, V_1(\sigma)$$

The proof of this theorem is contained in Ref. [28]. The theorem was used in Ref. [31] for constructing the Lyapunov function for a single machine system with flux decay effects.

We shall now proceed to construct Lyapunov function for the power system problem in the multi-machine case.

For the system to be stable, $\underline{N}$ and $\underline{Q}$ have to be chosen so that

$$\underline{Z}(s) = (\underline{N}+\underline{Q}s)W(s) \tag{4.119}$$

is positive real.

$\underline{N}$ is chosen as $\underline{N} = \underline{0}$ for Condition (ii) to be satisfied because $\underline{F}(\underline{\sigma})^T\underline{N}\underline{\sigma}$ can have negative values around the origin for a positive diagonal matrix $\underline{N}$. The function $V_1(\underline{\sigma})$ is chosen as

$$\begin{aligned} V_1(\underline{\sigma}) &= \sum_{k=1}^{m} \int_0^{\sigma_k} f_{1k}(\underline{\sigma})\,d\underline{\sigma} \\ &= \sum_{i=1}^{n-1} \sum_{j=i+1}^{n} B_{ij}[E_iE_j(\cos\delta_{ij}^s-\cos\delta_{ij}) \\ &\quad - (\delta_{ij}-\delta_{ij}^s)E_i^sE_j^s\sin\delta_{ij}^s] \end{aligned} \tag{4.120}$$

The partial derivatives of $V_1(\underline{\sigma})$ are written as follows:

$$\frac{\partial V_1}{\partial \sigma_k} = B_{ij}[E_iE_j\sin(\sigma_k+\delta_{ij}^s) - E_i^sE_j^s\sin\delta_{ij}^s] = f_{1k}(\underline{\sigma})$$

$$\text{for } k=1,2,\ldots,m \tag{4.121}$$

$$\frac{\partial V_1}{\partial \sigma_k} = \sum_{j=1}^{n} B_{ij}E_j(\cos\delta_{ij}^s-\cos\delta_{ij}) = f_{2i}(\underline{\sigma})$$

$$\text{for } k=m+1,\ldots,m+n \tag{4.122}$$

Combining both, we get

$$\nabla\underline{V}_1(\underline{\sigma}) = \underline{F}(\underline{\sigma}) \tag{4.123}$$

Hence $\underline{Q} = \underline{I}$.

To solve for the $\underline{P}$ matrix, we basically follow the same procedure as in Sec. 4.7.2. In this instance, the $\underline{P}$ matrix is given by

$$\underline{P} = \begin{array}{c} \\ n \\ n-1 \\ n \end{array} \overset{\begin{array}{cccc} n & n-1 & n \end{array}}{\left[\begin{array}{ccc} \underline{P}_1 & \underline{P}_2^T & \underline{0} \\ \underline{P}_2 & \underline{P}_3 & \underline{0} \\ \underline{0} & \underline{0} & \underline{P}_4 \end{array}\right]} = \overset{\begin{array}{cc} 2n-1 & n \end{array}}{\left[\begin{array}{cc} \underline{P}_c & \underline{0} \\ \underline{0} & \underline{P}_4 \end{array}\right]}\begin{array}{c} 2n-1 \\ \\ n \end{array} \qquad (4.124)$$

With the choice of $\underline{N} = \underline{0}$ and $\underline{Q} = \underline{I}$, the matrix $\underline{Z}(s)$ is expressed as

$$\underline{Z}(s) = \left[\begin{array}{cc} \underline{K}^T(s\underline{I}+\underline{\lambda})^{-1}\underline{M}^{-1}\underline{K} & \underline{0} \\ \underline{0} & s(s\underline{I}+\underline{\alpha})^{-1}\underline{\beta} \end{array}\right]$$

$$= \left[\begin{array}{cc} \underline{Z}_1(s) & \underline{0} \\ \underline{0} & \underline{Z}_2(s) \end{array}\right] \qquad (4.125)$$

Since $\underline{Z}(s)$ is the direct sum of $\underline{Z}_1(s)$ and $\underline{Z}_2(s)$, the condition for $\underline{Z}(s)$ to be positive real can be tested independently for $\underline{Z}_1(s)$ and $\underline{Z}_2(s)$. Clearly, Condition (i) and (ii) of positive real property in Eq. (4.61) are satisfied for $\underline{Z}_1(s)$ and $\underline{Z}_2(s)$. For Property (iii) to be satisfied, it is sufficient to show that

$$Z_i(j\omega) + Z_i^*(-j\omega) \geq 0, \qquad i=1,2 \qquad (4.126)$$

After some algebraic manipulation, we obtain

$$\underline{Z}_1(j\omega) + \underline{Z}_1^*(-j\omega) = \underline{K}^T\left[\text{Diag}\left(\frac{2D_i}{M_i^2\omega^2 + D_i^2}\right)\right]\underline{K} \qquad (4.127)$$

$$\underline{Z}_2(j\omega) + \underline{Z}_2^*(-j\omega) = \left[\text{Diag}\left(\frac{2\beta_i\omega^2}{(\omega^2+\alpha_i^2)}\right)\right] \qquad (4.128)$$

It is verified that both (4.127) and (4.128) are positive semi-definite. Therefore $\underline{Z}_1(s)$ and $\underline{Z}_2(s)$ are positive real which then guarantees that $\underline{Z}(s)$ is also positive real.

The solution for elements of $\underline{P}$ matrix in (4.124) using Eqs. (4.68) - (4.70) proceeds along the same lines as Sec. 4.7.2. Because of the structure of $\underline{P}$, we can solve for $\underline{P}_c$ as for the classical model with no transfer conductances. Any of the quadratic forms in the Lyapunov functions (4.91) - (4.97) can be chosen.

The solution for $\underline{P}_4$ corresponding to $\underline{W}_2(s)$ is obtained as

$$\underline{P}_2 = \underline{\alpha}\ \underline{\beta}^{-1} \tag{4.129}$$

Hence, the Lyapunov function for the whole system is

$$V(\underline{x}) = \frac{1}{2}\,\underline{x}_1^T\,\underline{P}_c\,\underline{x}_1 + \frac{1}{2}\,\Delta\underline{E}^T\,\underline{\alpha}\,\underline{\beta}^{-1}\,\Delta\underline{E}$$

$$+ \sum_{k=1}^{m} \int_0^{\sigma_k} f_{1k}(u_k)\,du_k$$

where $\Delta\underline{E} = [E_1 - E_1^s, \ldots, E_n - E_n^s]^T$

$$\underline{x}_1 = [\omega_1, \ldots, \omega_n \,\vert\, \delta_{1n} - \delta_{1n}^s, \ldots, \delta_{n-1,n} - \delta_{n-1,n}^s]^T$$

A special case of uniform damping can be discussed along similar lines since it affects the quadratic part only.

Conclusion

In this chapter, an attempt has been made to derive Lyapunov functions in a systematic manner both for single and multi-machine systems using many of the techniques discussed in Chapter II. For multimachine systems, we focussed on the energy type Lyapunov function which was derived using the COA reference frame and the general type Lyapunov function using the multivariable Popov Criterion. The latter is seen to be more general and can be extended to systems containing flux decay effects as well.

References

1. Willems, J. L., "Direct Methods for Transient Stability Studies in Power System Analysis", IEEE Trans. Automatic Control, Vol. AC-16, No. 4, August 1971, pp. 332-341.

2. Dharma Rao, N., "Generation of Lyapunov Functions for the Transient Stability Problem", Trans. Eng. Inst., Canada, Vol. 11, Rep. C-3, October 1968.

3. Peczkowski, J. L. and Liu, R. W., "A Format Method for Generating Lyapunov Functions", Journal of Basic Engineering, ASME, June 1967, pp. 433-439.

4. Mansour, M., "Stability Analysis and Control of Power Systems", in: Handschin, E. (Editor) "Real Time Control of Electric Power Systems" (Book), Elsevier Publishing Co., Amsterdam, 1972.

5. Yu, Y. N. and Vongsuriya, K., "Nonlinear Power System Stability Study by Liapunov Function and Zubov's Method", IEEE Trans., Vol. PAS-16, No. 12, Dec. 1967, pp. 1480-85.

6. De Sarkar, A. K. and Dharma Rao, N., "Zubov's Method and Transient Stability Problems of Power Systems", Proc. IEE, Vol. 118, No. 8, August 1971, pp. 1035-1040.

7. Pai, M. A., Mohan, M. A. and Rao, J. G., "Power System Transient Stability Regions Using Popov's Method", IEEE Trans., Vol. PAS-89, 1970, pp. 788-794.

8. Walker, J. A. and McLamroch, N. H., "Finite Regions of Attraction for the Problem of Lure", Int. J. Control, Vol. 6, No. 4, 1967, pp. 331-336.

9. Aylett, P. D., "The Energy-Integral Criterion of Transient Stability Limits of Power Systems", Proc. IEE, Vol. 105C, 1958, pp. 527-536.

10. Athay, T., Podmore, R. and Virmani, S., "A Practical Method for Direct Analysis of Transient Stability", IEEE Trans., Vol. PAS-98, No. 2, March/April 1979, pp. 573-584.

11. El-Abiad, A. H. and Nagappan, K., "Transient Stability Region of Multi-machine Power Systems", IEEE Trans., Vol. PAS-85, 1966, pp. 169-179.

12. Willems, J. L., "Optimum Lyapunov Functions and Stability Regions of Multimachine Power Systems", Proc. IEE, Vol. 117, No. 3, March 1970, pp. 573-578.

13. Ribbens-Pavella, M., "Critical Survey of Transient Stability Studies of Multimachine Power Systems by Lyapunov's Direct Method", Proc. Ninth Allerton Conference on Circuit and System Theory, October 6-8, 1971.

14. Pai, M. A., "Transient Stability Studies in Power Systems Using Lyapunov Popov Approach", Proc. 5th World IFAC Congress, Paris, 1972, Paper 31.5.

15. Anderson, B.D.O., "A System Theory Criterion for Positive Real Matrices", SIAM Journal (Control), 1967, Vol. 5, pp. 171-182.

16. Pai, M. A. and Murthy, P. G., "New Lyapunov Functions for Power Systems Based on Minimal Realizations", Int. J. Control, 1974, Vol. 19, No. 2, pp. 401-405.

17. Willems, J. L. and Willems, J. C., "The Application of Lyapunov Methods to the Computation of Transient Stability Regions for Multimachine Power Systems", IEEE Trans., Vol. PAS-89, 1970, pp. 795-801.

18. Ribbens-Pavella, M. and Howard, J. L., "Practical Considerations on the Transient Stability Power Systems Studies by Lyapunov's Direct Method", Revue E, Vol. VII, No. 5, 1973, pp. 167-169.

19. Willems, J. L., "A Partial Stability Approach to the Problem of Transient Power System Stability", Int. J. Control, Vol. 19, No. 1, 1974, pp. 1-14.

20. Henner, V. E., "A Multimachine Power System Lyapunov Function Using the Generalized Popov Criterion", Int. J. Control, Vol. 19, No. 5, 1974, pp. 969-976.

21. Pai, M. A. and Murthy, P. G., "On Lyapunov Functions for Power Systems with Transfer Conductances", IEEE Trans. Automatic Control, Vol. AC-18, No. 2, April 1973, pp. 181-183.

22. Uemura, K., Matuski, J., Yamada, J. and Tsuji, T., "Approximation of an Energy Function in Transient Stability Analysis of Power Systems", Electrical Engineering in Japan, Vol. 92, No. 6, 1972.

23. Gudaru, U., "A General Lyapunov Function for Multimachine Power Systems with Transfer Conductances", Int. J. Control, Vol. 21, No. 2, 1975, pp. 333-343.

24. Willems, J. L., Comments on Ref. [23], Int. J. Control, Vol. 23, No. 1, 1976, pp. 147-148.

25. Pai, M. A. and Varvandkar, S. D., "On the Inclusion of Transfer Conductances in Lyapunov Functions for Multimachine Power Systems", IEEE Trans. Automatic Control, Vol. AC-22, No. 6, 1977, pp. 983-985.

26. Ribbens-Pavella, M., "Transient Stability of Multimachine Power Systems by Lyapunov's Direct Method", Paper 71-CP17-PWR, IEEE Winter Power Meeting, New York, Jan. 1971.

27. Kakimoto, N., Ohsawa, Y. and Hayashi, M., "Transient Stability Analysis of Electric Power System via Luré Type Lyapunov Function", Parts I and II, Trans. IEE of Japan, Vol. 98, No. 5/6, May/June 1978.

28. Kakimoto, N., Ohsawa, Y. and Hayashi, M., "Transient Stability Analysis of Multimachine Power Systems with Field Flux Decays via Lyapunov's Direct Method", IEEE Trans., Vol. PAS-99, No. 5, Sept./Oct. 1980, pp. 1819-1827.

29. Desoer, C. A. and Wu, M. Y., "Stability of a Nonlinear Time-Invariant Feedback System Under Almost Constant Inputs", Automatica, Vol. 5, 1969, pp. 231-233.

30. Di Caprio, V. and Saccomano, F., "Nonlinear Stability Analysis of Multimachine Electric Power Systems", Richerche Di Automatica, Vol. 1, No. 1, Sept. 1970, pp. 1-29.

31. Pai, M. A. and Rai, V., "Lyapunov-Popov Stability Analysis of Synchronous Machine with Flux Decay and Voltage Regulator", Int. Journal of Control, Vol. 20, No. 2, Aug. 1974, pp. 203-212.

Chapter V

COMPUTATION OF STABILITY REGIONS FOR MULTIMACHINE POWER SYSTEMS

5.1 Introduction

The computation of t_{cr} for a given fault using Lyapunov's method involves three distinct steps,

(i) Constructing a $V(\underline{x})$ for the final post-fault stable system.

(ii) Constructing a region of attraction around the post-fault equilibrium state defined by $V(\underline{x}) < V_{cr}$.

(iii) Integration of the faulted system equations until $V(\underline{x}) = V_{cr}$ which then yields t_{cr}.

While step (i) has been discussed in Chapters III and IV, step (iii) is generally a straightforward exercise in numerical integration. It is, however, step (ii) which has been the biggest impediment to the successful practical application of Lyapunov's method ever since the method was first proposed in 1966 for computing t_{cr} directly instead of by repetitive simulation [1]. For the past decade and a half, considerable research effort has been devoted to this problem. In general, it was found that t_{cr} computed by Lyapunov's method was much less than the actual t_{cr} although for certain faults the results were fairly accurate. In view of this inconsistency in results, there was not much enthusiasm on the part of the power industry to make use of this technique either as an aid for simulation or for transient security assessment purposes. Fortunately, the picture is quite optimistic now due to some recent breakthroughs on the research front [2,3,4]. In this chapter, we shall discuss these new techniques in detail while briefly alluding to earlier efforts which gave conservative results. However, the formalism of the earlier methods is still useful in explaining the new methods. It is interesting to observe that the new approaches were possible precisely because of a deeper insight into the concept of region of attraction via the notion of invariant sets [5].

5.2 Methods for Computing Stability Regions in Power Systems

Theorem 2.7.5 on the region of asymptotic stability, which is pertinent to our discussion, is repeated below in a slightly modified form [6].

Theorem 5.1. Consider the autonomous system

$$\dot{\underline{x}} = \underline{F}(\underline{x}), \ \underline{F}(\underline{0}) = 0 \tag{5.1}$$

Let $V(\underline{x})$ be a scalar function. Suppose that the region $R = \{\underline{x} | V(\underline{x}) < k\}$ is bounded. Let $\dot{V}(\underline{x})$ be the derivative of $V(\underline{x})$ along the solution of (5.1). If $V(\underline{x})$ is positive definite and $\dot{V}(\underline{x})$ is negative definite in R, then the origin is asymptotically stable and all motions originating in R converge to the origin as $t \to \infty$.

The assumption on negative definiteness of $\dot{V}(\underline{x})$ in R can be relaxed to the condition: $\dot{V}(\underline{x}) \leq 0$ in R and does not vanish identically along any solution of (5.1) in R except the null solution.

5.2.1 Theoretical Approaches

(i) A practical method of estimating k was proposed as follows [6]: Let $V(\underline{x})$ be positive definite and $\dot{V}(\underline{x})$ be negative definite near the origin. Let V_{cr} be the lowest value of $V(\underline{x})$ on the surface $\dot{V}(\underline{x}) = 0$. Then the bounded region $R = \{\underline{x} | V(\underline{x}) < V_{cr}\}$ about the origin is a region of asymptotic stability and is contained in the exact domain of attraction. Zubov's method discussed in Chapter II (Sec. 2.14) and applied to a single-machine infinite bus system in Chapter IV (Sec. 4.5) precisely uses this formulation. The method, however, is unattractive from a computational point of view for higher order systems since it would result in a large nonlinear minimization problem.

(ii) A second method is to generalize the results of Walker and McLamroch [7], as applied to the single machine case in Sec. 4.6, for the multi-nonlinear case and apply it to the multi-machine system [8].

In Sec. 4.7.2, it has been shown that the multivariable Popov criterion is applicable only to systems containing negligible transfer conductances, since only then that the nonlinearities satisfy a sector condition around the origin. The state model for nonuniform damping case and with zero transfer conductances is given by (4.58) as

$$\begin{aligned} \dot{\underline{x}} &= \underline{A}\underline{x} - \underline{B}\underline{f}(\underline{\sigma}) \\ \underline{\sigma} &= \underline{C}\underline{x} \end{aligned} \tag{5.2}$$

The nonlinearity $\underline{f}_k(\underline{\sigma}) = f_k(\sigma_k)$ satisfies the sector condition $\sigma_k f_k(\sigma_k) > 0$ $k=1,2,\ldots,m$ $(m = n(n-1)/2)$ in the region

$$\ell_{k2} < \sigma_k < \ell_{k1}, \quad \ell_{k1} > 0, \quad \ell_{k2} < 0 \tag{5.3}$$

The Lyapunov function using the multivariable Popov theorem has been shown to be of the form

$$\begin{aligned} V(\underline{x}) &= \underline{x}^T \underline{P}\underline{x} + 2 \int_{\underline{0}}^{\underline{\sigma}} \underline{f}^T(\underline{\sigma}) \underline{Q} d\underline{\sigma} \\ &= \underline{x}^T \underline{P}\underline{x} + 2 \sum_{i=1}^{m} \int_0^{\sigma_i} q_i f_i(\sigma_i) d\sigma_i \end{aligned} \tag{5.4}$$

Define the polytope surrounding the origin as: $\{\underline{\sigma}:\sigma_k \subset (\ell_{k2},\ell_{k1})\}$. The boundary of this polytope is a set of 2m hyperplanes defined by

$$\sigma_k = \begin{cases} \ell_{k1} \\ \\ \ell_{k2} \end{cases} \qquad k=1,2,\ldots,m \tag{5.5}$$

A one-shot minimization of $V(\underline{x})$ on the boundary of this polytope will again result in a large number of nonlinear algebraic equations which can be computationally burdensome [15]. An alternative is to consider one hyperplane at a time and then choose the minimum of $V(\underline{x})$ over all such hyperplanes [8].

Even then the computational efforts are not reduced substantially. However, the mathematical formulation underlying this approach leads to an approximate method of computing V_{cr} which is related to the physical aspects of the problem, namely, the modes of instability. It is also related to the method using the unstable equilibrium points which is briefly described next.

(iii) It is well known that surrounding the origin (post-fault SEP) there are a large number of unstable equilibrium points satisfying the post-fault equilibrium equations

$$\begin{aligned} \underline{0} &= \underline{A}\underline{x} - \underline{B}\underline{f}(\underline{C}\underline{x}) \\ \text{i.e.} \quad \underline{0} &= \underline{F}(\underline{x}) \end{aligned} \tag{5.6}$$

We will interpret (5.6) in terms of the physical variables to get a better insight.

In terms of δ_i's and ω_i's, the model corresponding to (5.2) is given by (3.17) as

$$\begin{aligned} \dot{\omega}_i &= (-D_i/M_i)\omega_i + P_i - P_{ei} \qquad i=1,2,\ldots,n \quad \text{(a)} \\ \dot{\delta}_{in} &= \omega_{in} \qquad i=1,2,\ldots,n-1 \quad \text{(b)} \end{aligned} \tag{5.7}$$

The equilibrium solutions of the post-fault system are obtained by setting $\dot{\omega}_i = 0$ and $\dot{\delta}_{in} = 0$ in (5.7). This implies $\omega_i = \omega_n$. In most transient stability studies, the assumption $\sum_{i=1}^{n} (P_i - P_{ei}) = \sum_{i=1}^{n} P_i = 0$ is made since $\sum_{i=1}^{n} P_{ei} = 0$ in the absence of transfer conductances. Since $P_i = P_{mi} - |E_i|^2 G_{ii}$, we have

$$\sum_{i=1}^{n} P_{mi} - \sum_{i=1}^{n} |E_i|^2 G_{ii} = 0 \tag{5.8}$$

Since ω_i's represent the speed deviations from the reference speed ω_o, it is easily shown that the condition (5.8) can be met by changing the reference speed without affecting equations (5.7). If we add all the equations in (5.7a), we get

$$\sum_{i=1}^{n} \dot{\omega}_i + \sum_{i=1}^{n} \left(-\frac{D_i}{M_i}\right)\omega_i = 0$$

This implies that $\omega_i \to 0$ on $t \to \infty$. Hence, the equilibrium points of the post-fault system are obtained from the equations

$$\omega_i = 0; \quad P_i - P_{ei} = 0 \qquad i=1,2,\ldots,n \tag{5.9}$$

Although (5.9) represents n equations in (n-1) variables only (angle differences) they are not overdetermined in view of (5.8). Hence, one of the equations in (5.9) is redundant and the equations to be solved are

$$P_i - P_{ei} = 0 \qquad i=1,2,\ldots,n-1 \tag{5.10}$$

In the case of uniform or zero damping, it may be shown that the equations are

$$\begin{aligned} &\omega_i = 0 && i=1,2,\ldots,n \\ &\frac{(P_i - P_{ei})}{M_i} - \frac{(P_n - P_{en})}{M_n} = 0 && i=1,2,\ldots,n-1 \end{aligned} \tag{5.11}$$

Let δ_i^s represent the stable solution of (5.10) or (5.11). Generally, δ_i^s is close to the prefault values δ_i^o. In addition to $(\delta_i^s, \omega_i = 0)$ i=1,2,...,n (or equivalently the origin $\underline{x} = 0$ in (5.7)), there are a large number of other solutions of (5.10) or (5.11) which represent unstable equilibrium points (u.e.p.'s). The exact number of such u.e.p.'s surrounding the origin has been a matter of much speculation and discussion [9-12]. However, it is not relevant to our discussion regarding the region of stability. Inside the polytope (5.5) we have $V(\underline{x}) > 0$ and $\dot{V}(x) \leq 0$ and also origin is the only equilibrium point inside the polytope. The u.e.p.'s therefore must either lie on the boundary or outside the polytope. This conclusion is also consistent with the fact that Lyapunov's method yields only sufficient conditions. Since $\underline{F}(\underline{x}) = \underline{0}$ at the u.e.p.'s, $\dot{V}(\underline{x}) = (\underline{\nabla V})^T \underline{F}(\underline{x}) = 0$ at each of the u.e.p.'s. Hence, $V(\underline{x})$ can be computed at these u.e.p.'s and the minimum of these values can be designated as V_{cr}. This therefore provides the third method of estimating the region of attraction. In fact, this was precisely the method suggested in Ref. [1] in 1966. However, it was quickly realized that computation of these u.e.p.'s was not an easy task. Two of the most widely used methods to compute the u.e.p.'s from (5.10) or (5.11) were the Davidon-Fletcher-Powell (DFP) Method [9] and the Newton-Raphson method [10,11,12]. Each of these methods takes considerable computer time and the fact that $V(\underline{x})$ has to be computed at the large number of u.e.p.'s makes the evaluation of V_{cr} again computationally unattractive. However, through a combination of mathematical and physical reasoning, it can be shown that the u.e.p.'s of interest lie in the proximity of certain points on the boundary of the polytope defined by (5.5), and called as approximate u.e.p.'s. These points are easily identifiable and can be associated with the physical modes of loss of synchronism. This also leads to the derivation of the number of possible u.e.p.'s as $2(2^{n-1}-1)$.

5.2.2 Approximate u.e.p.'s and Modes of Instability [8,13]

Consider the system (5.2) and a typical hyperplane defined by

$$\sigma_k = \underline{c}_k\underline{x} = \ell_{k1} \tag{5.12}$$

where $\underline{c}_k = (c_{k1},\ldots,c_{k,2n-1})$ is the kth row of $\underline{C}$. We are seeking a minimum of $V(\underline{x})$ on this hyperplane. This implies that (5.9) is tangential to $V(\underline{x})$ at that point. The conditions for tangency are

$$\frac{\partial}{\partial x_j} V(\underline{x}) = \lambda_k \frac{\partial}{\partial x_j}(\underline{c}_k\underline{x}-\ell_{k1}) \qquad j=1,2,\ldots,2n-1 \tag{5.13}$$

where λ_k is a constant.

Assuming $V(\underline{x})$ is to be of the form (5.4) and after some algebraic manipulation, (5.13) becomes

$$2\underline{P}\underline{x} + 2\underline{C}^T\underline{Q}\underline{f}(\underline{\sigma}) - \underline{C}^T\underline{\lambda} = \underline{0} \tag{5.14}$$

where $\underline{\lambda} = (0,\ldots,\lambda_k,\ldots,0)^T$, $\underline{Q} = \text{Diag}(q_i)$. With $\underline{\sigma} = \underline{C}\underline{x}$ substituted in (5.14), (5.12) and (5.14) together constitute a set 2n equations in λ_k and the 2n-1 unknowns $\underline{x}$. Let the solutions be denoted by $\hat{\underline{x}}$ and $\hat{\underline{\sigma}}$. The minimum value of $V(\underline{x})$ can be shown to be [8]

$$V^k_{min} = \underline{\hat{\sigma}}^T[(\underline{C}\underline{P}^{-1}\underline{C}^T)^{-1}]^T\underline{\hat{\sigma}} + 2\int_0^{\underline{\hat{\sigma}}} \underline{f}^T(\underline{\sigma})\underline{Q}d\underline{\sigma} \tag{5.15}$$

Since there are 2m hyperplanes corresponding to (5.5), the minimum of $V(\underline{x})$ over all these hyperplanes gives an estimate of V_{cr}.

The preceding theory is general enough for any physical system described by (5.1). In the case of the power system model, it may be verified that not all σ_k's are linearly independent. In fact, only (n-1) of the $m = (\frac{n(n-1)}{2})$ σ_k's are linearly independent and the rest (m-n+1) variables can be obtained as a linear combination of the (n-1) σ_k's. Thus $\underline{\sigma}$ can be partitioned as

$$\underline{\sigma} = \begin{bmatrix} \underline{C}_I \\ \underline{C}_D \end{bmatrix} \underline{x}$$

where the subscripts I and D refer to linearly independent and dependent σ_k's.

A convenient choice of $\underline{\sigma}_I$ is $[\delta_{in}-\delta^s_{in},\ldots,\delta_{n-1,n}-\delta^s_{n-1,n}]$. The 2(n-1) hyperplanes are

$$\left.\begin{aligned} \sigma_k &= \pi - 2\delta^s_{kn} \\ \sigma_k &= -\pi - 2\delta^s_{kn} \end{aligned}\right\} \quad k=1,2,\ldots,n-1 \tag{5.16}$$

The polytope defined by (5.16) now has 2(n-1) hyperplanes and it is sufficient to find the minimum of $V(\underline{x})$ over these 2(n-1) instead of the 2m hyperplanes [17].

The case of single nonlinearity can now be considered as a particular case of the above formulation when $\underline{\sigma} = \sigma$, a scalar. Then

$$\hat{\sigma} = \text{Min}(|\ell_2|,|\ell_1|) = (\pi-2\delta^s)$$

and

$$V_{cr} = V_{min}(\underline{x}) = \frac{(\pi-2\delta^s)^2}{[\underline{c}\underline{P}^{-1}\underline{c}^T]} + q\int_0^{(\pi-2\delta^s)} f(\sigma)d\sigma \tag{5.17}$$

where $\underline{c}$ is now a row vector. This result is identical to (4.46) obtained in Chapter IV.

In the case of power systems, the boundary of the polytope defined by (5.16) in the angle space corresponds approximately to the potential energy boundary surface [5] on which the term corresponding to the integral in the V-function in (5.4) has a relative maximum (potential energy term). The quadratic term representing the kinetic energy has a relative minimum and is also much smaller in absolute value compared to the integral term near the tangency point of $V(\underline{x})$ to the hyperplane. Thus, we can neglect the quadratic term in comparison to the integral term in $V(\underline{x})$. Equation (5.14) becomes

$$\underline{C}^T(\underline{Q}\underline{f}(\underline{\sigma}) - \frac{1}{2}\underline{\lambda}) = \underline{0} \tag{5.18}$$

Since $\underline{C} = [\overset{n}{\underline{0}} \quad \overset{n-1}{\underline{S}}]$, Eq. (5.18) can be simplified as

$$\underline{S}^T(\underline{Q}\underline{f}(\underline{\sigma}) - \frac{1}{2}\underline{\lambda}) \tag{5.19}$$

Equation (5.19) is a set of (n-1) equations. Since only (n-1) of the σ_i's are linearly independent, the other (m-n+1) σ_i's can be expressed as a linear combination of the (n-1) σ_i's. Thus, in Eq. (5.19) we will have (n-1) equations with the unknowns $(\sigma_1, \sigma_2, \ldots, \sigma_{k-1}, \sigma_{k+1}, \ldots, \sigma_{n-1})$ and λ_k. A solution to Eq. (5.13) which satisfies the restrictions on other σ_i's $(i \neq k)$ is the tangency point. The minimization can also be performed on the intersection of more than one hyperplane at a time.

The structure of Eq. (5.19) will be explained for the case of a 4-machine system. $\underline{C}$ is given from (3.47) as

$$\underline{C} = \left[\begin{array}{c|ccc} \underline{0} & & \underline{I} & \\ \hline & 1 & -1 & 0 \\ \underline{0} & 1 & 0 & -1 \\ & 0 & 1 & -1 \end{array}\right] \tag{5.20}$$

Choose the linearly independent set of σ_i's as $\sigma_1 = \delta_{14} - \delta_{14}^s$, $\sigma_2 = \delta_{24} - \delta_{24}^s$, $\sigma_3 = \delta_{34} - \delta_{34}^s$. This implies choosing machine

4 as a reference machine. We have $\sigma_4 = \sigma_1 - \sigma_2$, $\sigma_5 = \sigma_1 - \sigma_3$, $\sigma_6 = \sigma_2 - \sigma_3$. The polytope with respect to σ_1, σ_2, σ_3 is shown in Fig. 5.1. Let the hyperplane on which minimization is to be done be $\sigma_2 = \ell_{21} = \pi - 2\delta_{24}^s$. Then Eq. (5.19) becomes

$$q_1 f_1(\sigma_1) + q_4 f_4(\sigma_1 - \ell_{21}) + q_5 f_5(\sigma_1 - \sigma_3) = 0$$

$$- q_4 f_4(\sigma_1 - \ell_{21}) + q_6 f_6(\ell_{21} - \sigma_3) = \lambda_2/2 \qquad (5.21)$$

$$q_3 f_3(\sigma_3) - q_5 f_5(\sigma_1 - \sigma_3) - q_6 f_6(\ell_{21} - \sigma_3) = 0$$

The above equations are to be solved for σ_1, σ_3 and λ_2 such that σ_1 and σ_3 satisfy $\ell_{12} < \sigma_1 < \ell_{11}$, $\ell_{32} < \sigma_3 < \ell_{31}$. These are nonlinear equations and can be solved by any of the well known methods for solving a system of nonlinear algebraic equations. However, one needs good initial guesses (starting points) and these are obtained by associating physical modes of instability with points on the polytope. Thus, for the

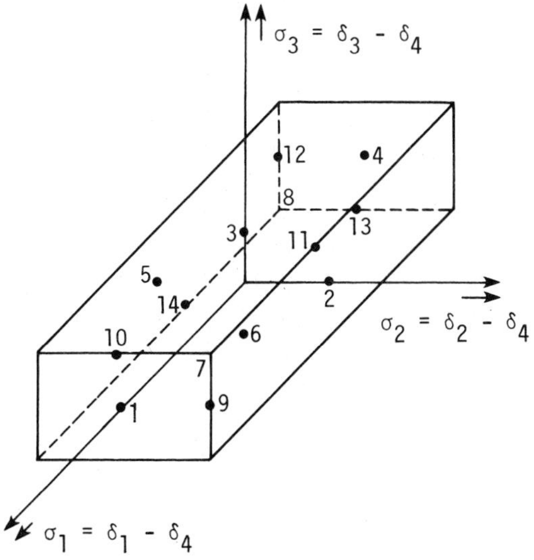

Fig. 5.1. Polytope for a 4-machine system with machine 4 as reference.

minimization on the hyperplane $\sigma_2 = \pi - 2\delta_{24}^s$, we can take $\sigma_1 = \sigma_3 = 0$ which is equivalent to assuming that machine 2 is losing synchronism by acceleration with respect to the reference machine while other machines are at their post-fault SEP's. This initial guess is also now an approximate unstable equilibrium point (a.u.e.p.). In the polytope of Fig. 5.1, this point is 2, i.e. $\sigma_2 = \ell_{21}$ and $\sigma_1 = \sigma_3 = 0$. If we had taken the hyperplane $\sigma_2 = \ell_{22} = -\pi - 2\delta_{24}^s$, then the a.u.e.p. would be $\sigma_2 = \ell_{22}$, and $\sigma_1 = \sigma_3 = 0$, i.e. point 5 on the polytope. Physically, this point corresponds to machine 2 decelerating with respect to machine 4 while other machines are stable. Similarly, we can take the other hyperplanes and thus generate in all 6 approximate u.e.p.'s numbered 1-6 corresponding to one machine losing synchronism either by acceleration or deceleration. To this we must add two other a.u.e.p.'s on the polytope, one corresponding to all the three machines accelerating with respect to the reference, i.e. $\sigma_1 = \ell_{11}$, $\sigma_2 = \ell_{21}$, $\sigma_3 = \ell_{31}$ (point 7 in Fig. 5.1) and the other corresponding to all the three machines decelerating with respect to reference machine, i.e. $\sigma_1 = -\ell_{12}$, $\sigma_2 = -\ell_{22}$, $\sigma_3 = -\ell_{32}$ (point 8 in Fig. 5.1). It is easy to verify that these two points correspond electrically to machine 4 losing synchronism by itself. In all, we have eight physical modes of instability of one machine losing synchronism. For an n-machine case, these will be $2({}^{n-1}C_1 + {}^{n-1}C_{n-1}) = 2n$.

We next take two machines going out of step at a time either by accelerating or decelerating together. Suppose we choose the hyperplanes

$$\sigma_2 = \ell_{21} = \pi - 2\delta_{24}^s$$

$$\sigma_3 = \ell_{31} = \pi - 2\delta_{34}^s$$

Then following the same formalism we will get three equations but the unknowns will be σ_1 and the two constants λ_2 and λ_3. Physically, this corresponds to machines 2 and 3 accelerating and hence a starting point to obtain a solution of the equations is $\sigma_2 = \pi - 2\delta_{24}^s$, $\sigma_3 = \pi - \delta_{34}^s$, $\sigma_1 = 0$. This is point 9 on the polytope in Fig. 5.1. In a 4-machine case, there are a total of six possibilities of two machines accelerating or decelerating together, i.e. $2({}^{n-1}C_2)$. These are points 9-14 in Fig. 5.1. We already have considered three machines accelerating or decelerating together (points 7 and 8). Thus, in a 4-machine case, we have in all 14 modes of instabilities or a.u.e.p.'s of the accelerating or decelerating type. These can be summarized into the following types:

(a) Each machine going out of step with respect to the reference machine by acceleration

$$(\pi-2\delta_{14}^s,0,0),\ (0,\pi-2\delta_{24}^s,0),\ (0,0,\pi-2\delta_{34}^s) \tag{5.22a}$$

(b) Two machines going out of step at a time by accelerating

$$(\pi-2\delta_{14}^s,\ \pi-2\delta_{24}^s,\ 0),\ (\pi-2\delta_{24}^s,\ 0,\ \pi-2\delta_{34}^s),\ (0,\ \pi-2\delta_{24}^s,\ \pi-2\delta_{34}^s) \tag{5.22b}$$

(c) All the three machines going out of step at a time by accelerating

$$(\pi-2\delta_{14}^s,\ \pi-2\delta_{24}^s,\ \pi-2\delta_{34}^s) \tag{5.22c}$$

(d) Repeat (a), (b), (c) with $+\pi$ replaced by $-\pi$ i.e. loss of synchronism by deceleration instead of acceleration. (5.22d)

In the case of n machines, we can generalize the result as follows: There are ${}^{(n-1)}C_k$ possibilities of k machines going out of step by acceleration and an equal number by deceleration. We can denote the approximate u.e.p. as

$$(\underbrace{\pi-2\delta_{in}^s,}_{k\ \text{terms}}\quad \underbrace{0,0,\ldots}_{(n-1-k)}) \tag{5.23}$$

for instability by acceleration and with $+\pi$ replaced by $-\pi$ for deceleration. Thus, the total number of modes of instabilities and hence the corresponding number of u.e.p.'s are

$$2\sum_{k=1}^{n-1} {}^{n-1}C_k = 2({}^{n-1}C_1 + {}^{n-1}C_2 + \ldots {}^{n-1}C_{n-1}) = 2(2^{n-1}-1)$$

For large n, this number can be very large [9]. Computation of such a large number of u.e.p.'s is not practical and hence there have been efforts to limit the search to a credible number of u.e.p.'s. This is discussed next.

5.3 Quick Determination of V_{cr} Through Approximate u.e.p.'s [9,13]

It has been the practical experience that generally whenever the system loses synchronism one of the machines will pull out first either by accelerating or decelerating. Hence, it was proposed that the search for u.e.p.'s be limited to those in the proximity of the following approximate u.e.p.'s:

(i) (n-1) of the type $(0,0,\ldots\pi-2\delta^s_{in},0\ldots0)$ $i=1,2,\ldots,n-1$.

(ii) (n-1) of the type (i) with $+\pi$ replaced by $-\pi$.

(iii) one of the type $(\pi-2\delta^s_{1n},\pi-2\delta^s_{2n},\ldots,\pi-2\delta^s_{n-1,n})$.

(iv) one of the type (iii) with $+\pi$ replaced by $-\pi$.

Thus in all there are 2n approximate u.e.p.'s. It has been shown by Prabhakara and El-Abiad [11] mathematically and practically validated by Ribbens-Pavella [9] that the changes in $V(\underline{x})$ in the neighborhood of an unstable equilibrium point are very small. Thus, evaluating $V(\underline{x})$ at these 2n approximate u.e.p.'s gives a good approximation to the value of $V(\underline{x})$ at the neighboring u.e.p.'s. A practical method therefore would be

1. Compute $V(\underline{x})$ at the 2n approximate u.e.p.'s.
2. Choose among the 2n approximate u.e.p.'s the one which gives minimum value of $V(\underline{x})$. Let this u.e.p. be designated as $\underline{x}^u_{appr}$.
3. Using Newton-Raphson or Davidon-Fletcher-Powell method, compute $\underline{x}^u$ with $\underline{x}^u_{appr}$ as the starting point from (5.10) or (5.11).
4. $V_{cr} = V(\underline{x}^u)$.

This method has given satisfactory results on relatively small sized systems. An improvement over this method is to limit the search to n points as follows: [14]

(i) Compute the sign of $P_kM - M_kP$ (k=1,2,...,n) where $P = \Sigma P_k$ and denote it by S_k.

(ii) The n approximate u.e.p.'s are

$$(0,\ldots,(S_k\pi-2\delta^s_{kn}),\ldots,0) \qquad k=1,\ldots,n-1)$$

$$(-S_n\pi-2\delta^s_{1n},\ldots,-S_n\pi-2\delta^s_{n-1,n})$$

In the methods discussed so far, it has been observed that a certain degree of conservatism in the computation of t_{cr} appears as the size of the system increases. The reasons for this as well as methods which overcome this basic deficiency are discussed in the rest of this chapter.

5.4 New Approaches to Transient Stability Investigation

A basic lacuna in the method based on evaluating V_{cr} as the V_{min} at the u.e.p.'s or approximate u.e.p.'s lies in postulating the mode of instability based on post-fault configuration only. This is incorrect as the following reasoning shows. For the time being, assume that the post-fault configuration is the same for the the different facult locations, i.e. there is no line switching. The use of same value of V_{min} implies that irrespective of the fault location, the same machine/group of machines goes unstable corresponding to the u.e.p. which determined V_{min}. Physically, this is not true and we know that the actual mode of instability is influenced by the fault location, the type of fault and the fault clearing time. In general, the critically cleared but unstable trajectories for different fault locations will exhibit different modes of instability and hence will pass through or in the vicinity of different u.e.p.'s. This is especially true as the system size increases. In small systems with a tight interconnection, faults at different locations may result in the same machine going unstable. Any new approach to the problem must therefore take into account the faulted trajectory and the mode of instability it induces in the post-fault system. Hence, computation of V_{cr} becomes fault dependent and for each fault there is a relevant or a controlling u.e.p. V_{cr} should be the value of V at this relevant u.e.p. for the particular fault. There are three approaches to the problem and we shall discuss all three of them in varying detail. All of them have a fast method of computing the faulted trajectory through some approximation. The methods are:

(i) Method based on acceleration criteria and determining the controlling u.e.p. through a two-stage process [4].

(ii) Method based on computing the controlling u.e.p. through a Davidon-Fletcher-Powell algorithm after getting a good approximation for it using the faulted trajectory and principle singular surface properties [3,5].

(iii) Method based on the notion of a potential energy boundary surface surrounding the post-fault SEP and use of the faulted trajectory to compute V_{cr} directly [2,5].

5.4.1 Approximate u.e.p. and Acceleration Criteria [4]

In Ref. [1] which presented Lyapunov's method for stability analysis for the first time, it was observed that the machine having the largest ratio of initial acceleration to inertia constant at $t = 0^+$ would be expected to accelerate faster than the other machines. Then it is probable that this machine would become unstable first. Hence, the approximate u.e.p. to be chosen would be

$$(0,\ldots,\pi-2\delta_{kn}^{s},\ldots,0)$$

where k is the machine having the largest ratio of acceleration to inertia. This conjecture may work well with small systems but is likely to give inconsistent results in larger systems. In Ref. [4] the accelerations are computed not $t = 0^+$, but a little ahead on the faulted trajectory by which time the pattern of instability will have been established. This method incorporates the procedure outlined in Sec. 5.3 as the first conservative estimate of t_{cr} and then updates both the likely mode of instability and the t_{cr}. The steps to be followed are:

1. Compute the value $V_{cr,appr}$ as the minimum of $V(\underline{x})$ at all the 2n a.u.e.p.'s as in Sec. 5.3.
2. Execute forward numerical integration of the faulted system. At each time step compare $V(\underline{x})$ with $V_{cr,appr}$. Let t_{cL} correspond to the case when $V(\underline{x}) = V_{cr,appr}$.
3. Compute the accelerations of all the machines at t_{cL}. Let k be the machine with the largest acceleration at $t = t_{cL}$. This machine may or may not be the same as in step 1. Corresponding to machine k, find the value of V(x) at the a.u.e.p. $(0,\ldots,\pi-2\delta_{kn}^{s},\ldots,0)$. This can be taken to be V_{cr} for all practical purposes. If necessary, the exact u.e.p. $\underline{x}^u$ can be computed using the a.u.e.p. as the starting point and V_{cr} can be corrected.
4. Continue numerical integration until $V(\underline{x}) = V_{cr}$. This method gives a good estimate of actual t_{cr}.

What the above method needs is a fast numerical method to integrate the faulted equations from $t = 0$ to $t = t_{cL}$. In fact, numerical integration can be avoided altogether by having a Taylor Series expansion around the prefault steady state solution [14].

$$\delta_i(t_e) = \sum_p \delta_i^{(p)}(t_o) \frac{(t_e-t_o)^p}{p!} \qquad i=1,2,\ldots,n \tag{5.24}$$

t_o is the initial instant when the disturbance occurs.

t_e denotes the time when the system reaches its next configuration (for instance, when the fault is cleared).

$\delta_i^{(p)}$ is the pth derivative of $\delta_i(t)$.

For clearing times not exceeding 0.5-0.6 sec, experience shows that the Taylor series expansion gives good agreement with the actual trajectories by taking p = 4. Thus

$$\delta_i(t_e) = \sum_{p=0}^{4} \delta_i^{(p)}(t_o^+) \frac{(t_e-t_o)^p}{p!} \qquad i=1,2,\ldots,n \qquad (5.25)$$

The four derivatives are obtained as follows:

$$\left.\begin{aligned} \delta_i(t_o^+) &= \delta_i(t_o^-) \\ \delta_i^{(1)}(t_o^+) &= \delta_i^{(1)}(t_o^-) \end{aligned}\right\} \qquad i=1,2,\ldots,n \qquad (5.26)$$

since there is no instantaneous change in rotor angles or relative speeds at t = 0. From the swing equation (neglecting damping)

$$M_i \frac{d^2\delta_i}{dt^2} = P_{mi} - P_{ei}(\underline{\delta}) \qquad i=1,2,\ldots,n \qquad (5.27)$$

$$\delta_i^{(2)}(t_o^+) = \frac{1}{M_i}(P_{mi}-P_{ei}(\underline{\delta}(0^+)) \qquad i=1,2,\ldots,n \qquad (5.28)$$

$$\delta_i^{(p)}(t_o^+) = -\frac{1}{M_i} P_{ei}^{(p-2)}(t_o^+) \qquad i=1,2,\ldots,n \qquad (5.29)$$

For p = 4, $P_{ei}(\underline{\delta})$ therefore has to be differentiated twice to obtain $\delta_i^{(3)}(t_o^+)$ and $\delta_i^{(4)}(t_o^+)$.

5.4.2 Numerical Results [4]

The above method has been tested on a large number of test systems such as 3, 7, 9, 15 and 40 machine system with satisfactory results. The results for the 7 machine CIGRE system is presented here. Figure 5.2 shows the single line diagram of the seven machine system and Table 5.1 gives the data for the system. The values of t_{cr} obtained by this method are compared with the standard Lyapunov criterion (i.e. V_{cr} computed as V_{min} at the u.e.p.'s) as well as the step-by-step method and are shown in Table 5.2. Column 1 of the table indicates the bus where the fault is applied. In column 2, I indicates that the post-fault configuration is the same as prefault configuration. D indicates that the two configurations differ because of line switching. It is observed that while the standard Lyapunov criterion gives consistently

GENERATORS

Bus	P_{base} (MVA)	X (%) ([1])	M (MW s^2/rad)	P_m (MW)	E (p.u)	δ^o (°)
1	100	7.4	6.02	217	1.106	7.9
2	100	11.8	4.11	120	1.156	-0.2
3	100	6.2	7.59	256	1.098	6.5
4	100	4.9	9.54	300	1.110	3.9
5	100	7.4	6.02	230	1.118	7.0
6	100	7.1	6.77	160	1.039	3.6
7	100	8.7	5.68	174	1.054	7.9

LOADS

Bus	P (MW)	Q (MVar)	Bus	P (MW)	Q (Mvar)
2	200	120	8	100	50
4	650	405	9	230	140
6	80	30	10	90	45
7	90	40			

LINES

Bus	R (ohm)	X (ohm)	ωC/2 (μS)
1 - 3	5	24.5	200
1 - 4	5	24.5	100
2 - 3	22.8	62.6	200
2 - 10	8.3	32.3	300
3 - 4	6	39.5	300
3 - 9	5.8	28	200
4 - 5	2	10	200
4 - 6	3.8	10	1200
4 - 9	24.7	97	200
4 - 10	8.3	33	300
6 - 8	9.5	31.8	200
7 - 8	6	39.5	300
8 - 9	24.7	97	200

([1]) These values include the transformer's reactances and are expressed on a 100 MVA base.

Table 5.1 Data of the CIGRE 7-machine system (Reproduced from Ref. [14])

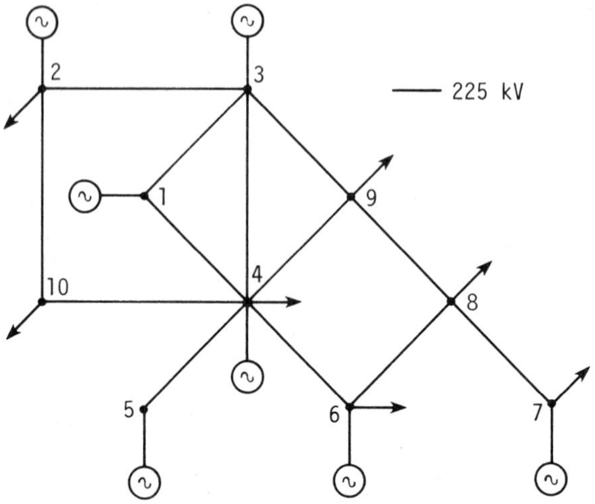

Fig. 5.2. 7 machine CIGRE test system.

conservative estimates, the present method gives excellent agreement with actual t_{cr} except for a conservative estimate for fault at bus #2 and a pessimistic estimate at bus #8.

Fault Location	Final Configuration	Step-by-Step $t_{cr}(10^{-2}s)$	Standard Lyapunov Method $t_{cr}(10^{-2}s)$	Acceleration Criteria Method $t_{cr}(10^{-2}s)$
1	I	35-36	26-27	35-36
2	I	41-42	34-35	36-37
3	I	39-40	25-28	39-40
4	I	50-51	30-31	47-48
5	I	35-36	26-27	35-36
6	I	52-53	39-40	50-51
8	D	44-45	40-41	50-51
6	D	51-52	34-35	50-51

Table 5.2 Critical clearing time by acceleration criterion and standard Lyapunov method for the 7-machine system

The computing time by this method compared to the step-by-step method (with three simulations) is improved by a factor of 100 [4].

5.4.3 Transient Energy Analysis Method

The reason for conservativeness of Lyapunov's method was systematically examined in Ref. [5] and they proposed a methodology which takes into account the dynamics of the faulted system and also the transfer conductance terms. The concept of a controlling u.e.p. was introduced and discussed extensively in the context of both first swing instability and multiswing instabilities. The notion of a controlling u.e.p. for first swing instability is also contained in Ref. [2]. While the former uses the center of angle reference formulation, the latter uses the machine angle reference frame. The two methods, however, differ in the way the transfer conductance terms are accounted for. Reference [5] provides a proper rationale for taking into account the faulted dynamics via the concept of invariant sets in Lyapunov stability theory. Both approaches treat Lyapunov function as an energy function comprising the sum of kinetic and potential energies and make use of the properties of the potential energy function.

Invariant set and region of attraction

Consider a set of disturbances (faults) $i=1,2,\ldots,\ell$ at different locations which, after being cleared, will result in the same post-fault configuration. Typically these will be faults at the generator/load buses which are self-clearing without any line switching. For conceptual clarity, we divide the system dynamics into two phases as in Eq. (3.14) and (3.15).

Phase A (Faulted phase) $\dot{\underline{x}} = \underline{F}_1(\underline{x},\underline{P}_i^f)$, $\underline{x}(0) = \underline{x}_0$, $0 < t \leq t_c$
where $\underline{P}_i^f$ is the parameter vector different for each ith fault. Typically, the elements of this parameter vector are the G_{ij}, B_{ij} elements in the swing equation.

Phase B (Post-fault phase) $\dot{\underline{x}} = \underline{F}_2(\underline{x},\underline{P}^{pf})$ $t \geq t_c$.

The solutions of the equation $\underline{0} = \underline{F}_2(\underline{x},\underline{P}^{pf})$ are the equilibrium points of the post-fault system in the state space. Let $\underline{x}_e$ be the SEP and the remaining equilibrium points will be the u.e.p.'s.

Basically, the transition from the state $\underline{x}_o$ to $\underline{x}_e$ is what is of interest to the power engineer. The two sets of d.e's

$$\begin{aligned} \dot{\underline{x}} &= F_1(\underline{x},\underline{P}_i^f), \ \underline{x}(0) = \underline{x}_o & 0 < t \leq t_c \\ \dot{\underline{x}} &= F_2(\underline{x},\underline{P}^{pf}), \ \underline{x}(t_c) = \underline{x}_{t_c} & t \geq t_c \end{aligned} \tag{5.30}$$

form a discontinuous set of d.e's describing the motion of the state variables. Given a certain t_c, if the state of the system goes from $\underline{x}_o$ to $\underline{x}_e$, the system is stable. The maximum t_c for which this is possible is called t_{cr}. Because of the inherent mathematical difficulties associated with stability investigation of a set of discontinuous d.e's via Lyapunov's method, we consider only the post-fault d.e and construct a region of stability around $\underline{x}_e$.

Definition

An invariant set for a system of equations $\dot{\underline{x}} = \underline{f}(\underline{x})$, $\underline{x}(0) = \underline{x}_o$ is defined as any subset of the state space having the property that a trajectory starting at $t = t_o$ in Ω belongs to Ω for all $t \geq t_o$. For example, the equilibrium state $\underline{x}_e$ is an invariant set and so is any trajectory of $\dot{\underline{x}} = \underline{f}(\underline{x})$.

In the power system problem, we define the invariant set for the post-fault system as the set of all trajectories of the post-fault system

$$\dot{\underline{x}} = \underline{F}_2(\underline{x},\underline{P}^{pf}), \quad t \geq t_c \tag{5.31}$$

whose initial conditions $\underline{x}(t_c)$ lie on the faulted trajectory

$$\dot{\underline{x}} = \underline{F}_1(\underline{x},\underline{P}_i^f), \quad \underline{x}(0) = \underline{x}_o \tag{5.32}$$

for the fault i.

In Fig. 5.3, the faulted trajectory is shown by the thick line and the invariant set by the dotted trajectories.

The region of attraction is defined for the post-fault SEP, $\underline{x}_e$ with respect to this invariant set. The following theorem is directly applicable.

Theorem 5.1

Let Ω be an invariant set for $\dot{\underline{x}} = \underline{F}_2(\underline{x}, \underline{P}_i^{pf})$, $t > t_c$ with $\underline{x}_e$ belonging to Ω. Let $V(\underline{x}-\underline{x}_e)$ be defined on Ω with $V(\underline{0}) = 0$. Let s(k) denote the set

$$s(k) = \{\underline{x} \varepsilon \Omega \,|\, V(\underline{x}-\underline{x}_e) < k\}$$

Suppose for some $k_o > 0$, that

(i) $V(\underline{x}-\underline{x}_e)$ is positive definite and decrescent on $s(k_o)$, and

(ii) $\dot{V}(\underline{x}-\underline{x}_e)$ is negative definite on $s(k_o)$

then $s(k_o)$ is contained in the region of attraction of $\underline{x}_e$.

We shall now interpret this theorem for the power system problem. Along the faulted trajectory for i^{th} fault $V(\underline{x}-\underline{x}_e)$ computed for the post-fault system increases with increasing t. For $t < t_{cr}$, the trajectories of the post-fault system converge to $\underline{x}_e$ and for this subset of the invariant set $V(\underline{x}-\underline{x}_e)$ is positive definite and $\dot{V}(\underline{x}-\underline{x}_e)$ is negative definite. For $t > t_{cr}$, the post-fault trajectories diverge away from $\underline{x}_e$ and hence $\dot{V}(\underline{x}-\underline{x}_e)$ is not negative definite. Consequently, for this subset of the invariant set there does not exist a valid Lyapunov function. The value k_o is the value of $V(\underline{x})$ at $t = t_{cr}$ on the faulted trajectory. For $t = t_{cr}$, the post-fault trajectory becomes unstable and theoretically passes through but practically goes in the vicinity of $\underline{x}^{ui}$, the controlling u.e.p. for fault i. This is analogous to the single machine case where the critical trajectory for the post-fault system lies on the separatrix. The critical trajectory is shown hatched in Fig. 5.4. For another fault location $\underline{P}_j^f$, we have a different controlling u.e.p. $\underline{x}^{uj}$ (also shown hatched). The computation of $k_o = V_{cr}$ for each fault is

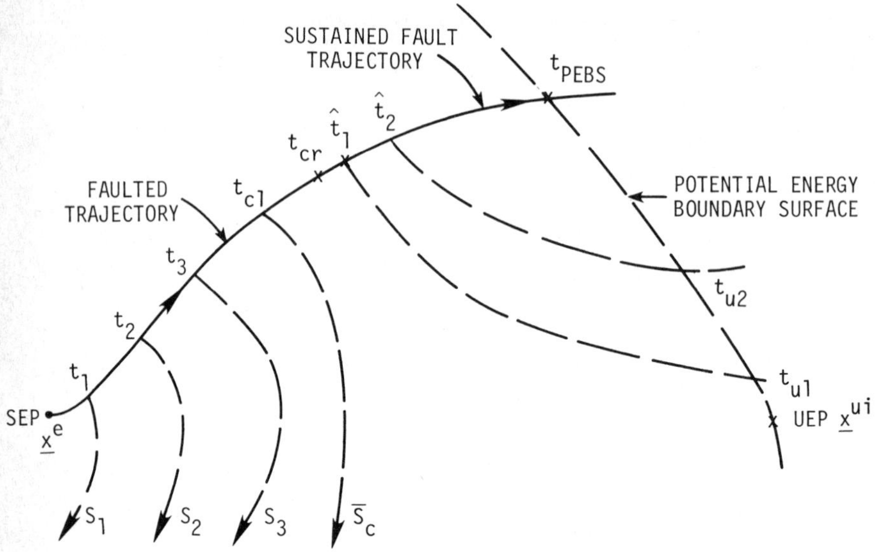

Fig. 5.3. Illustration of the faulted and post fault trajectories for various clearing times corresponding to stable and unstable cases.

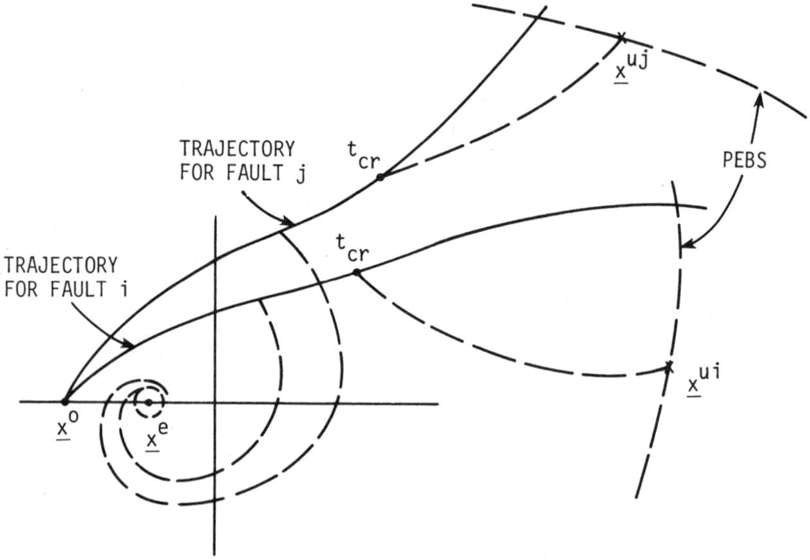

Fig. 5.4. Invariant sets and critical trajectories for different faults.

therefore the crux of the problem even though the post-fault configuration is identical. If the system is conservative, $V(\underline{x})$ remains constant after the fault is cleared and hence the value of k_o at $t = t_{cr}$ is same as the value of $V(\underline{x})$ at $\underline{x} = \underline{x}^{u_i}$ where $\underline{x}^{u_i}$ is the <u>controlling</u> u.e.p. All research work in Lyapunov stability analysis of power systems until 1978 considered the value of k_o as the minimum of $V(\underline{x})$ evaluated at each of the $\underline{x}^{u_i}$ irrespective of fault location for the same post-fault configuration. It is therefore clear why the results obtained by such a procedure were conservative in terms of t_{cr}. Since k_o has to be computed for each fault condition, the algorithm must emphasize computational speed. Methods (ii) and (iii) cited at the beginning of Sec. 5.4 which achieve this will now be discussed.

5.4.4 Method Based on Computing Controlling u.e.p. Through DFP Routine [3]

The center of angle formulation is used. The equations are (neglecting damping)

$$M_i \dot{\tilde{\omega}}_i = P_i - P_{ei} - \frac{M_i}{M_T} P_{COA} \triangleq f_i(\underline{\theta}) \tag{5.33}$$

$$\dot{\theta}_i = \tilde{\omega}_i \qquad i=1,2,\ldots,n$$

The Lyapunov function is given by

$$\begin{aligned} V(\underline{\theta},\underline{\tilde{\omega}}) &= \frac{1}{2}\sum_{i=1}^{n} M_i\tilde{\omega}_i^2 - \sum_{i=1}^{n} P_i(\theta_i-\theta_i^s) \\ &\quad - \sum_{i=1}^{n}\sum_{j=i+1}^{n} C_{ij}(\cos\theta_{ij}-\cos\theta_{ij}^s) \\ &\quad - \int_{\theta_i^s+\theta_j^s}^{\theta_i+\theta_j} D_{ij}\cos\theta_{ij}\,d(\theta_i+\theta_j) \end{aligned} \tag{5.34}$$

The terms in this Lyapunov function (also referred to as the transient energy function) can be physically interpreted in the following way

(i) $$\frac{1}{2}\sum_{i=1}^{n} M_i\tilde{\omega}_i^2 = \frac{1}{2}\sum_{i=1}^{n} M_i\omega_i^2 - \frac{1}{2}M_T\omega_o^2 \tag{5.35}$$

Total change in rotor kinetic energy relative to COA = Total change in rotor kinetic energy minus the change in COA kinetic energy

(ii) $$\sum_{i=1}^{n} P_i(\theta_i-\theta_i^s) = \sum_{i=1}^{n} P_i(\delta_i-\delta_i^s) - \sum_{i=1}^{n} P_i(\delta_o-\delta_o^s) \tag{5.36}$$

Change in rotor potential energy relative to COA = Change in rotor potential energy minus the change in COA potential energy

(iii) $$C_{ij}(\cos\theta_{ij}-\cos\theta_{ij}^s) \tag{5.37}$$

Change in magnetic stored energy in branch ij

(iv) $$\int_{\theta_i^s+\theta_j^s}^{\theta_i+\theta_j} D_{ij}\cos\theta_{ij}\,d(\theta_i+\theta_j) \tag{5.38}$$

Change in dissipated energy in branch ij

The kinetic energy term (5.35) is denoted by $V_k(\underline{\tilde{\omega}})$, the sum of the potential energy terms (5.36) and (5.37) is denoted by $V_p(\underline{\theta})$ and the dissipative energy term (5.38) denoted by $V_d(\underline{\theta})$.

To successfully compute the u.e.p.'s in the presence of transfer conductances, the center of angle (COA) formulation is advantageous. The u.e.p.'s are the solutions of the equations obtained by setting $\dot{\tilde{\omega}}_i = \dot{\theta}_i = 0$ in (5.33). This gives

$$f_i(\underline{\theta}) = P_i - P_{ei} - \frac{M_i}{M_T}P_{COA} = 0$$
$$\tilde{\omega}_i = 0 \qquad i=1,2,\ldots,n \tag{5.39}$$

Since P_{ei} and P_{COA} are functions of differences of θ_{ij}'s only and $\theta_n = -1/M_n \sum_{i=1}^{n} M_i\theta_i$, it is sufficient to consider only the first (n-1) equations in (5.39) namely

$$f_i(\underline{\theta}) = P_i - P_{ei} - \frac{M_i}{M_T} P_{COA} \qquad i=1,2,\ldots,n-1 \qquad (5.40)$$

Equation (5.40) has several solutions, one of them being the post-fault stable equilibrium point (SEP) $\underline{\theta}^s$. Surrounding this are the unstable equilibrium points. At all the equilibrium points (stable or unstable) $\tilde{\omega}_i = 0$. There are two methods to solve the equation (5.40). One is the Newton-Raphson method which is discussed adequately in the literature [11,12]. The second method is by minimization of a scalar objection function $F(\underline{\theta}) = \sum_{i=1}^{n} f_i^2(\underline{\theta})$ which is physically the total bus power mismatch. The presence of the term $\frac{M_i}{M_T} P_{COA}$ in each equation ensures that the transmission power loss is not absorbed by the reference machine alone but by all machines in proportion to their inertias. The minimization is done using the DFP technique. While Newton-Raphson method is suited well for computing the stable equilibrium point, it has convergence problems with unstable equilibrium points and hence the DFP technique is preferred [5].

The u.e.p. to be computed is the controlling u.e.p. for the given fault condition. The procedure consists of two parts.

(i) Calculate the fault trajectory of the system, i.e. integrate Eq. (5.33) with parameters corresponding to the faulted network and initial conditions at t = 0.

(ii) From the faulted trajectory, estimate the likely mode of instability and hence the u.e.p. We will elaborate upon each of these aspects.

<u>Calculation of faulted trajectory:</u> Since the purpose of this computation is only to know in which direction the faulted trajectory is moving, a very accurate numerical integration is not required. Hence, we approximate $f_i(\underline{\theta})$ as

$$f_i \cong \alpha_i + \beta_i \cos \eta t \qquad (i=1,2,\ldots,n) \qquad (5.41)$$

α_i, β_i and η are computed on the basis of two power flows. The first at the instant of fault application fixes vectors $\underline{\alpha}$ and $\underline{\beta}$ for a given frequency η. The second, along an approximate trajectory shortly after the fault, is used to compute η. The details are contained in Ref. [5]. The justification for this simplification is based on the results obtained by the approximate method and the exact simulation for the duration of the faulted period. We have already seen in Sec. 5.4.1 another way of approximating the faulted trajectory through a Taylor series expansion.

Estimating the controlling u.e.p.: In the Lyapunov function (5.34), the sum of $V_p(\underline{\theta})$ and $V_d(\underline{\theta})$ represents the negative sum of the first integral of the system accelerating powers, i.e.

$$V_{PE}(\underline{\theta}) \triangleq V_p(\underline{\theta}) + V_d(\underline{\theta}) = -\sum_{i=1}^{n} \int_{\theta_i^s}^{\theta_i} f_i(\underline{\theta})d\theta_i \qquad (5.42)$$

This is referred to as the potential energy function. In the local region around $\underline{\theta}^s$ in the angle space and referred to as the principle region, the potential energy function is convex. The closed boundary of this region is denoted as the principle singular surface [5,16]. The principle singular surface is the multi-dimensional analog of the region of dynamic stability which in a 2-machine system corresponds to the peak of the power transfer curve. In the n-dimensional case, if we define $\theta_{in} = \theta_i - \theta_n$ $(i=1,2,\ldots,n-1)$ as the relative rotor angles, then a polytope defined by $|\theta_{in}| \leq \pi/2$, $i=1,2,\ldots,n-1$ and $|\theta_i - \theta_j| \leq \pi/2$ $i,j=1,2,\ldots,n-1$ can be inscribed within the principle singular surface [18]. The faulted trajectory is integrated until it intersects the principle singular surface. To know this point of intersection, the quantity $F(\underline{\theta}) = \sum_{i=1}^{n} f_i^2(\underline{\theta})$ for the post-fault system is monitored and when $F(\underline{\theta})$ reaches a maximum, the angles $\underline{\theta}^{ss}$ are close to the boundary of the principle singular surface [5]. This conclusion can be justified by the following supplementary reasoning. From (5.42), in the absence of transfer conductances, the gradient of $V_p(\underline{\theta})$ is the negative of the vector of accelerating powers $f_i(\underline{\theta})$. Hence, $F(\underline{\theta})$ can be interpreted as the Euclidean norm squared of the gradient of the potential energy function. Hence on the boundary of the principle region, $F(\underline{\theta})$ is a maximum. Interestingly, if the Jacobian of $\underline{f}(\underline{\theta})$ is computed and its eigenvalues are examined, they are negative within the principle region and on the principle region boundary surface one of the eigenvalues becomes zero [18]. The various steps in the algorithm are summarized as follows [5]:

Step 1

For the given fault, determine the fault trajectory approximation parameters, α_i, β_i, i=1,2,...,n and η.

Step 2

Calculate the parameters of the post-fault matrix $\underline{Y}_{red}$ and compute $\underline{\theta}^s$ from (5.40) using the Newton-Raphson or DFP technique with the pre-fault equilibrium point as the initial guess.

Step 3

Using approximation of Step 1, find the time at which the (post-fault) power mismatch function $F(\underline{\theta})$ is maximum along the fault-on trajectory. The angles at this time are $\underline{\theta}^{ss}$; this is (very close to) the intersection point of the fault-on trajectory with the principle singular surface.

Step 4

Starting with $\underline{\theta}^{ss}$, a vector $\underline{\theta}^{ss} - \underline{\theta}^s$ is formed and normalized to form the direction vector $\underline{h}$.

Step 5

Solve the one dimensional minimization problem

$$\min F(\underline{\theta}(z)) \triangleq F(\underline{\theta}(z^*)) \qquad z > 0$$

where $\underline{\theta}(z) \triangleq \underline{\theta}^{ss} + z \cdot \underline{h}$ and $\underline{\theta}(z^*) \triangleq \hat{\underline{\theta}}^u$

Step 6

With $\hat{\underline{\theta}}^u$ as the starting point, solve (5.40) using the DFP minimization technique to obtain the unstable equilibrium point $\underline{\theta}^u$. $V(\underline{0},\underline{\theta}^u)$ gives V_{cr} for the given fault.

The following comments serve to clarify Steps (1)-(6). Steps (1) and (2) are self-explanatory. In Step (3), the point where $F(\underline{\theta})$ is maximum signifies that maximum synchronizing capability of the post-fault system is reached. Also by this time, effects of the faulted trajectory will have been established. The basic idea in Step (4) is to construct a direction vector which points in the direction of the u.e.p. corresponding to the fault case under study. $\underline{\theta}^s$ corresponds to the origin in the polytope defined by (5.5) and among the different choices possible for the direction vector, this direction is chosen from the point of view of computational simplicity and determining the first swing instability mode. In Step (5), the minimization of the power mismatch function along the direction vector, constructed in the previous step, defines $\hat{\underline{\theta}}_u$ which is the point on the vector closest to the controlling u.e.p. In Step (6) after computing the exact u.e.p. $\underline{\theta}^u$, V_{cr} is computed as $V_p(\underline{\theta}^u) + V_d(\underline{\theta}^u)$ since at an u.e.p. $\tilde{\omega}_i = 0$. $V_d(\underline{\theta}^u)$ is an integral which is path dependent and hence the following approximate formula is used to evaluate $V_d(\underline{\theta}^u)$. This is derived from the integral expression for $V_d(\underline{\theta})$ by assuming a linear path of integration from $\underline{\theta}^s$ to $\underline{\theta}^u$.

$$V_d(\underline{\theta}^u) = \sum_{i=1}^{n-1} \sum_{j=i+1}^{n} \frac{\theta_i^u + \theta_j^u - \theta_i^s - \theta_j^s}{\theta_i^u - \theta_j^u - \theta_i^s + \theta_j^s} [\sin\theta_{ij}^u - \sin\theta_{ij}^s] D_{ij}$$

A numerical example illustrating the method is presented in Sec. 5.5.

5.4.5 Potential Energy Boundary Surface (PEBS) Method

In the previous section, we have seen that for a particular fault condition, an accurate estimate of V_{cr} can be obtained by computing $V(\underline{x})$ at the controlling u.e.p. $\underline{x}^u$. Since at an u.e.p. $\tilde{\omega}_i = 0$, $V_{cr} \triangleq V_{PE}(\underline{\theta}^u) = V_p(\underline{\theta}^u) + V_d(\underline{\theta}^u)$ i.e. V_{cr} is the potential energy computed at the controlling u.e.p. $\underline{\theta}^u$ in the angle space. This suggests that we can map the stability region from the state space onto the angular subspace of θ_i's and consider $V_p(\underline{\theta}) + V_d(\underline{\theta}) \triangleq V_{PE}(\underline{\theta})$ in this space. The $V(\underline{x}) = k$ (k = constant) curves in the $\underline{x}$ space, i.e. $\tilde{\underline{\omega}}, \underline{\theta}$ space can be projected on to the $\underline{\theta}$ space and interpreted as the $V_{PE}(\underline{\theta})$ = constant curves. These are known as the equipotential curves. These curves are closed in a small region around the stable equilibrium point (SEP $\underline{\theta}^s$ but otherwise have the general shape shown in Fig. 5.5 for a 3-machine case. Surrounding the SEP are a number of u.e.p.'s $\underline{\theta}^u$. At some of the u.e.p.'s called the saddle points, (U_1 and U_2) the gradient of $\underline{\nabla V}_p \equiv 0$, i.e.

$$\frac{\partial V_p}{\partial \theta_1} = \frac{\partial V_p}{\partial \theta_2} = \cdots\cdot \frac{\partial V_p}{\partial \theta_n} = 0. \tag{5.43}$$

The equipotential curves around the saddle points are not closed curves. There are other u.e.p.'s (U_3) in addition to the saddle points around which the equipotential contours are closed curves. Another way to characterize the SEP, saddle point type u.e.p.'s and the other u.e.p.'s, is through the eigenvalues of the Jacobian of $\underline{f}(\underline{\theta})$ at these points. At the SEP, the eigenvalues are negative, at the saddle point there are both positive and negative eigenvalues and at the other u.e.p.'s the eigenvalues are positive [18]. In terms of the potential energy, the SEP $\underline{\theta}^s$ has the minimum potential energy (zero if the post-fault SEP has been transferred to the origin). At the saddle point type u.e.p.'s, the energies have a relative mini-max pattern and at the other u.e.p.'s the energies have a relative maximum. We consider the u.e.p.'s in an internal of $-\pi$ to $+\pi$ around $\underline{\theta}^s$. We can also interpret the equipotential curves in the relative rotor angle reference frame δ_{in} as in Fig. (5.6). S is the stable SEP $\underline{\delta}^s$ while U_1, U_2 and U_3 are the saddle point type u.e.p.'s and U_4 is a u.e.p. where the potential energy is a relative maximum. The curves 0_1, 0_2, and 0_3 are orthogonal to the equipotential curves and pass through the u.e.p.'s U_1, U_2, U_3 and U_4. Both in Figs. (5.5) and (5.6) the dotted curves joining the u.e.p.'s are collectively referred to as the potential energy boundary surface (PEBS).

The motion of the system trajectory can be projected on to the angular space and the loss of synchronism (instability) phenomena can be viewed in the context of the u.e.p.'s as

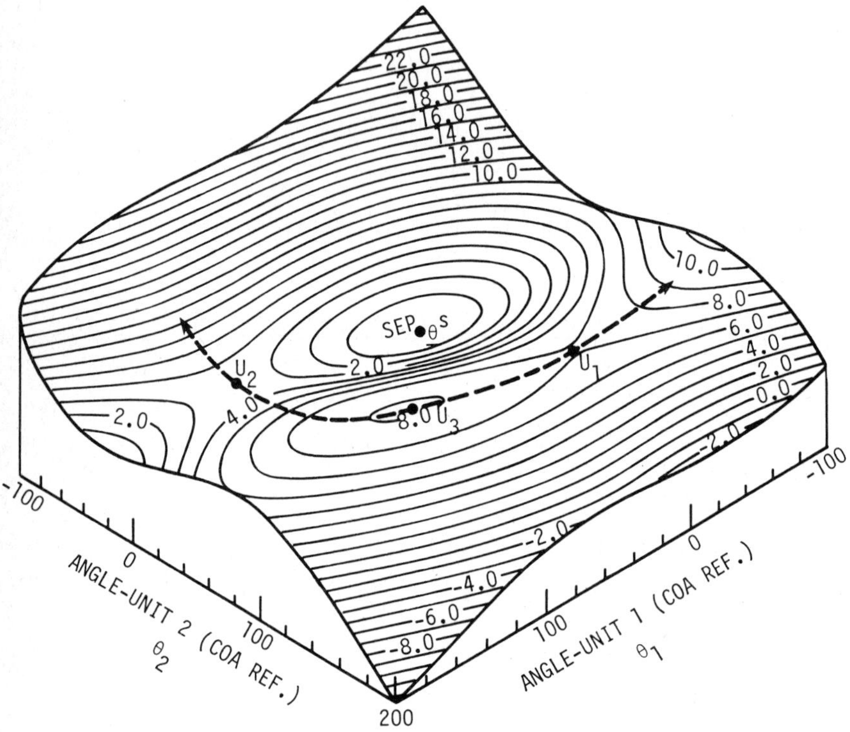

Fig. 5.5. Equi-potential energy surfaces (solid lines) and the potential energy boundary surface (dotted line) for three-machine system (reproduced from Ref. [5]).

follows (Figs. 5.4 and 5.7): If a fault is critically cleared and the system is unstable, then the trajectory will pass in the vicinity of one of the saddle point type u.e.p.'s. For a different location of the fault, the critically cleared but unstable trajectory will in general pass in the vicinity of another saddle point type u.e.p. even though the post-fault configuration may be the same for all the faults. Such a saddle point where the critically cleared trajectory escapes is termed as the controlling u.e.p. $\underline{\theta}^u$ for the given fault. It need not necessarily be the u.e.p. with the lowest energy as was assumed in all work prior to 1978 and which resulted in conservative estimates for t_{cr}. For $t > t_{cr}$ the trajectory will cross the PEBS at a point close to the relevant u.e.p. $\underline{\theta}^u$ if the system is first swing unstable. For multi-swing unstable cases, the unstable trajectory may cross the PEBS at an altogether different point on the PEBS. If the fault is cleared before t_{cr}, then the trajectory will oscillate within the PEBS and ultimately settle down at the SEP $\underline{\theta}^s$ if there is damping in the system.

If we restrict to the case of first swing unstable cases, then unstable trajectories including the sustained fault trajectory cross the PEBS at a point such that the following hold true (Fig. 5.7).

(i) The mismatch $F(\underline{\theta}) = \sum_{i=1}^{n} f_i^2(\theta)$ is close to zero indicating that the unstable trajectory is crossing the PEBS at a point close to the controlling u.e.p.

(ii) $V_{PE}(\underline{\theta})$ is maximum and $V_k(\tilde{\underline{\omega}})$ is minimum at the PEBS crossing.

It is therefore logical to assume as a good estimate of critical energy V_{cr} the value of $V_{PE}(\underline{\theta}) = V_p(\underline{\theta}) + V_d(\underline{\theta})$ at the PEBS crossing for the sustained trajectory. Experimental results indicate that the same procedure can be used for estimating V_{cr} even for multi-swing cases with satisfactory results [5]. In order to translate this procedure for practical purposes, we need a criterion to deted the PEBS crossing.

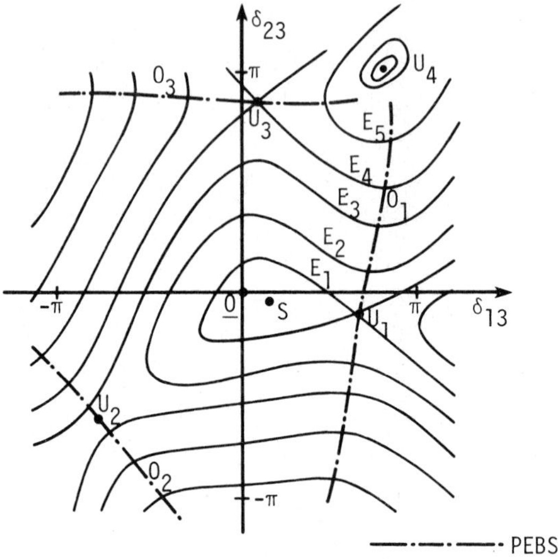

Fig. 5.6. Equipotential curves, PEBS and the UEP's in the relative rotor angle reference frame (adapted from Ref. [2]).

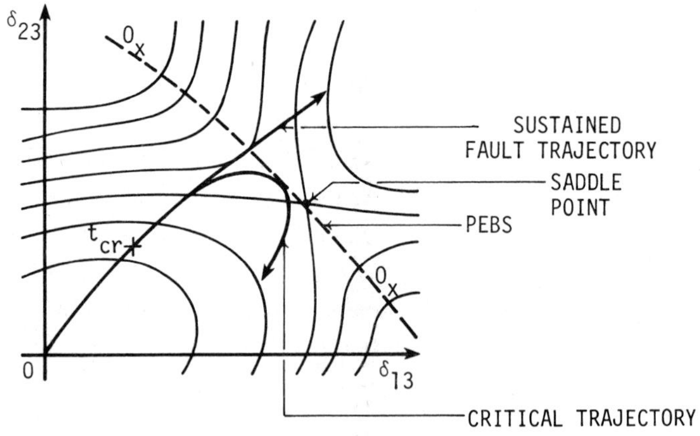

Fig. 5.7. PEBS crossing of critical and sustained fault trajectories (from Ref. [2]).

Characterization of PEBS

Consider the system (5.33) with negligible transfer conductances. The Lyapunov function (5.34) then is the sum of kinetic energy plus the potential energy.

$$V(\underline{\tilde{\omega}},\underline{\theta}) = V_k(\underline{\tilde{\omega}}) + V_p(\underline{\theta}) \tag{5.44}$$

$$\dot{V}(\underline{\tilde{\omega}},\underline{\theta}) = \dot{V}_k(\underline{\tilde{\omega}}) + \dot{V}_p(\underline{\theta})$$

$$= \sum_{i=1}^{n} (M_i\tilde{\omega}_i\dot{\tilde{\omega}}_i + \frac{\partial V_p}{\partial \theta_i}\dot{\theta}_i) \tag{5.45}$$

Define the vectors $\underline{T} = (M_1\dot{\omega}_1, M_2\dot{\omega}_2, \ldots, M_n\dot{\omega}_n)^T$ and

$$\frac{\partial V_p}{\partial \underline{\theta}} = (\frac{\partial V_p}{\partial \theta_1}, \frac{\partial V_p}{\partial \theta_2}, \ldots, \frac{\partial V_p}{\partial \theta_n})^T$$

Since $\dot{\theta}_i = \tilde{\omega}_i$, we have

$$\dot{V}(\underline{\tilde{\omega}},\underline{\theta}) = (\underline{T} + \frac{\partial V_p}{\partial \underline{\theta}})\cdot\underline{\tilde{\omega}} \tag{5.46}$$

Since the transfer conductances are negligible, the post-fault system is conservative and $\dot{V}(\underline{\tilde{\omega}},\underline{\theta}) = 0$. Hence for non-zero $\underline{\tilde{\omega}}$, we must have

$$\underline{T} = -\frac{\partial V_p}{\partial \underline{\theta}} \tag{5.47}$$

Therefore, the direction of the torque vector $\underline{T}$ is in a direction negative to the gradient of $V_p(\underline{\theta})$, i.e. $\underline{T} = -\underline{\nabla V}_p$. By definition $\underline{\nabla V}_p$ is in a direction of increasing $\underline{V}$ = constant contours in the angle space. Hence, the torque is always orthogonal to the equipotential surfaces and in the direction of decreasing V_p = constant contours.

Consider a trajectory crossing the PEBS. Inside the region surrounded by PEBS and containing the SEP, the projection of $\underline{T}$ on the velocity vector $\underline{\tilde{\omega}}$ is in a direction opposing the motion, i.e. towards the SEP. However, outside the PEBS, the projection of $\underline{T}$ is in the direction of the velocity vector. This suggests that the dot product of $\underline{\nabla V}_p$ with the velocity vector changes sign in the vicinity of the PEBS. In Fig. (5.8) a trajectory is shown crossing the PEBS such that the velocity vector is orthogonal to PEBS. It is easily verified that inside the PEBS the dot product $\underline{T}\cdot\underline{\tilde{\omega}} < 0$ and outside the PEBS, $\underline{T}\cdot\underline{\tilde{\omega}} > 0$ and on the PEBS, $\underline{T}\cdot\underline{\tilde{\omega}} = 0$. When the velocity vector does not cross the PEBS orthogonally, the point where $\underline{T}\cdot\underline{\tilde{\omega}} = 0$ does not lie on the PEBS. However, since $\underline{\nabla V}_p$ changes sign rapidly near PEBS, the change in sign occurs very close to the PEBS. In the case of a conservative system, the expression $\underline{T}\cdot\underline{\tilde{\omega}}$ can be simplified to

$$\underline{T}\cdot\underline{\tilde{\omega}} = \sum_{i=1}^{n} f_i(\underline{\theta})\cdot\tilde{\omega}_i = -\sum_{i=1}^{n} \frac{\partial V_p}{\partial \theta_i}\frac{d\theta_i}{dt} = -\dot{V}_p \qquad (5.48)$$

Hence inside the PEBS, $\dot{V}_p > 0$ and outside it is < 0.

Since $\dot{V}_p + \dot{V}_k = 0$ for a conservative system, we have $\dot{V}_k = -\dot{V}_p$. Hence, inside PEBS $\dot{V}_k < 0$ and outside the PEBS $\dot{V}_k > 0$. What these relations imply is that inside the stable region potential energy increases and kinetic energy decreases. Decrease in V_k means decrease in $|\omega_i|$ and prevents the system from getting outside the PEBS. The opposite is true outside of the PEBS. Consequently, when $\dot{V}_p$ or $\dot{V}_k$ changes sign can be interpreted to be the instant of time when the trajectory crosses the PEBS [2].

In the presence of transfer conductances, the above observations hold good only approximately. The reason is that although $V(\underline{\tilde{\omega}},\underline{\theta}) > 0$, $\dot{V}(\underline{\tilde{\omega}},\theta)$ is sign indefinite. The deviation of $\dot{V}(\underline{\tilde{\omega}},\underline{\theta})$ from zero is however not significant and if some damping is present $\dot{V}$ will be negative semidefinite [5]. There are two methods for monitoring the PEBS crossing for an unstable trajectory.

(i) Since $V(\underline{\tilde{\omega}},\underline{\theta}) = V_k(\underline{\tilde{\omega}}) + V_p(\underline{\theta}) + V_d(\underline{\theta})$ and since $V_k(\underline{\tilde{\omega}})$ is the same whether the transfer conductances are present or not, the instant of time when $\dot{V}_k = 0$ is taken as the PEBS crossing [2].

(ii) A more accurate method would be to have a constructive method to identify the PEBS [5].

Criterion	Inside PEBS	Outside PEBS	Comment
$\dot{V}_p$	Positive	Negative	zero transfer conductance case
$\dot{V}_k$	Negative	Positive	zero transfer conductance case
$\underline{f}^T(\underline{\theta})\cdot\hat{\underline{\theta}}$	Negative	Positive	Both zero and non-zero transfer conductance cases

Table 5.3 Summary of Criteria for PEBS Crossing

In Table 5.3, the criterion for crossing the PEBS based on the signs of $\dot{V}_p$ and $\dot{V}_k$ are summarized. When transfer conductances are neglected, either of these criteria can be used to compute t_{PEBS} (Fig. 5.3) on the sustained fault trajectory and $V_p(t_{PEBS})$ is an estimate of V_{cr} for that particular fault. Application of this criterion leads to difficulties when transfer conductances are present requiring a correction factor for V_{cr} [2]. However, another characterization of PEBS to be discussed next is valid for both zero and non-zero transfer conductance cases.

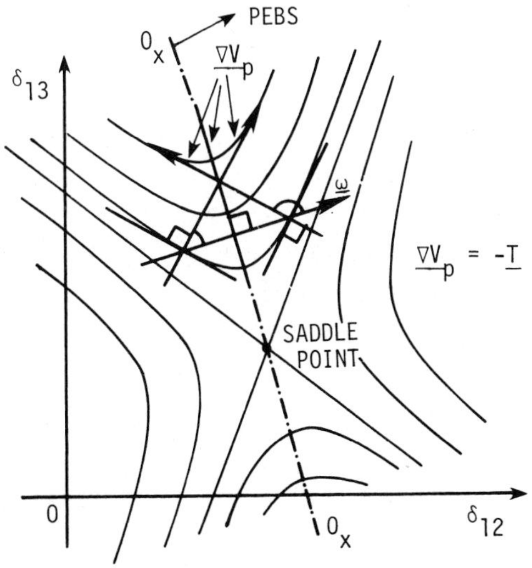

Fig. 5.8. The dot product of $\underline{\nabla V}_p$ with velocity vector in the vicinity of PEBS (from Ref. [2]).

Constructing the PEBS [5]

The nature of PEBS can be understood through a 3-machine example with transfer conductances neglected. Since only two θ_i's are linearly independent, $V_p(\underline{\theta})$ is shown in Fig. (5.5) in the space of θ_1 and θ_2. The post-fault configuration is the same as the pre-fault configuration so that there is only one SEP which also is the point of minimum potential energy. The two saddle points have energies of 3.2 and 7.2 respectively and the third u.e.p. has an energy of about 9.0. The dotted line joining the u.e.p.'s and orthogonal to the equipotential contours is the PEBS. The PEBS forms a "ridge" and can be defined by the following constructive procedure. Starting from the post-fault SEP, go out in every direction in the angle space. Along each direction find $\underline{\theta}^*$ where V_p attains a relative maximum. The set of points $\underline{\theta}^*$ found in the manner characterizes the PEBS.

(i) System with zero transfer conductances

Mathematically the PEBS can be defined by setting directional derivative of $V_p(\underline{\theta})$ to zero. By definition, the directional derivative of $V_p(\underline{\theta})$ is given by

$$\lim_{\alpha \to 0} \frac{V_p(\underline{\theta}^s+\alpha\underline{u}) - V_p(\underline{\theta}^s)}{\alpha}$$

where $\underline{u}$ is the unit vector along which the derivative is desired. In our case

$$\underline{u} = \frac{\underline{\theta} - \underline{\theta}^s}{||\underline{\theta} - \underline{\theta}^s||}$$

The directional derivative is equal to

$$\underline{\nabla V}_p(\underline{\theta})\cdot\underline{u} = \frac{1}{||\underline{\theta} - \underline{\theta}^s||} \sum_{i=1}^{n} \frac{\partial V_p}{\partial \theta_i} (\underline{\theta}_i - \underline{\theta}_i^s) \qquad (5.49)$$

Since on the PEBS, $V_p(\underline{\theta})$ has a relative maximum, we set the directional derivative to zero. Hence the PEBS is defined by $\underline{\theta} = \underline{\theta}^*$, satisfying

$$\sum_{i=1}^{n} \frac{\partial V_p}{\partial \theta_i} (\theta_i - \theta_i^s) = 0 \tag{5.50}$$

(ii) System with transfer conductances [5]

Let $f_i(\underline{\theta}) = P_i - P_{ei} - \frac{M_i}{M_T} P_{COA}$ = accelerating power of the ith machine. Then

$$V_p(\underline{\theta}) + V_d(\underline{\theta}) = -\sum_{i=1}^{n} \int_{\theta_i^s}^{\theta_i} f_i(\underline{\theta})\, d\theta_i \tag{5.51}$$

The linear angle path assumption in $\underline{\theta}$ space

$$\underline{\theta} = (\underline{\theta} - \underline{\theta}^s)\alpha + \underline{\theta}^s = \Delta\underline{\theta}\alpha + \underline{\theta}$$

leads to $V_p(\underline{\theta}) + V_d(\underline{\theta})$ being approximated as

$$V_{appr}(\underline{\theta}) = -\sum_{i=1}^{n} \int_0^{\alpha} f_i(\Delta\underline{\theta}\alpha + \underline{\theta}^s)\Delta\theta_i\, d\alpha$$

$$\frac{dV_{appr}}{d\underline{\theta}} = 0 \quad \text{implies} \quad -\sum_{i=1}^{n} f_i(\Delta\underline{\theta}\alpha + \underline{\theta}^s)\Delta\theta_i = 0 \tag{5.52}$$

If (5.52) is satisfied for some α, i.e., some $\underline{\theta}^*$, then it means that $\underline{\theta}^*$ satisfies

$$\sum_{i=1}^{n} f_i(\underline{\theta})(\theta_i - \theta_i^s) = 0 \tag{5.53}$$

This is the characterization of the PEBS when transfer conductances are present. In vector form, it can be rewritten as $\underline{f}^T(\underline{\theta})\cdot(\underline{\theta} - \underline{\theta}^s) = 0$. Denoting $\underline{\theta} - \underline{\theta}^s = \hat{\underline{\theta}}$, inside the PEBS, the quantity $\underline{f}^T(\underline{\theta})\cdot\hat{\underline{\theta}}$ is less than zero and outside the PEBS it is greater than zero. We can verify this statement by analogy with the zero transfer conductance case when $\underline{f}^T(\underline{\theta})\cdot\hat{\underline{\theta}}$ reduces $-(\frac{\partial V_p}{\partial \underline{\theta}})\cdot\hat{\theta}$ (since $\underline{f}(\underline{\theta}) = -\frac{\partial V_p}{\partial \underline{\theta}}$). Within the PEBS, when $\underline{\theta}$ is moving away from $\underline{\theta}^s$, both $\underline{\nabla V}_p$ and $\hat{\underline{\theta}}$ are > 0. When $\underline{\theta}$ is moving towards $\underline{\theta}^s$, both $\underline{\nabla V}_p$ and $\hat{\underline{\theta}}$ are < 0. Hence, inside the PEBS $\underline{\nabla V}_p\cdot\hat{\underline{\theta}} > 0$. By a similar reasoning outside the PEBS the dot product will be < 0 and on the PEBS equal to zero. Note the difference between this criterion and the criterion $\underline{T}\cdot\underline{\tilde{\omega}}$ for a conservative system given by (5.48). They do not in general give the same PEBS crossing.

Potential Energy Approximation in Systems with Transfer Conductances [5]

The energy function when transfer conductances are not neglected, contains an integral term which is path dependent, i.e.

$$V_d(\underline{\theta}) = \sum_{i=1}^{n-1} \sum_{j=i+1}^{n} \int_{\theta_i^s+\theta_j^s}^{\theta_i+\theta_j} D_{ij} \cos\theta_{ij}\, d(\theta_i+\theta_j)$$

$$= \sum_{i=1}^{n-1} \sum_{j=i+1}^{n} I_{ij} \tag{5.54}$$

These terms represent the energy dissipated in the transfer conductances. The correct path is that of the actual system trajectory.

When the trajectory is known, the terms in Eq. (5.54) above can be evaluated using the trapezoidal rule as follows.

$$\begin{aligned} I_{ij}(m) = I_{ij}(m-1) + \frac{1}{2} D_{ij}[&\cos(\theta_i(m)-\theta_j(m)) \\ &+ \cos(\theta_i(m-1)-\theta_j(m-1))] \\ &\cdot [\theta_i(m) + \theta_j(m) \\ &- \theta_i(m-1) - \theta_j(m-1)] \end{aligned}$$

where $I_{ij}(m)$ is the value at the mth step, $I_{ij}(m-1)$ at the $(m-1)^{th}$ step, $\theta_i(m)$, $\theta_i(m-1)$ and $\theta_j(m)$, $\theta_j(m-1)$ are the angles at bus i and j at the mth and the $(m-1)^{th}$ time steps respectively, and $I_{ij}(0) = 0$ for all i,j.

This potential energy approximation is used in computing $V_d(\underline{\theta})$ for the fault-on trajectory as well in the last step when t_{cr} is evaluated. This numerical approximation avoids the need to have an analytical expression for $V_d(\underline{\theta})$.

Computation of V_{cr}, t_{cr} by the PEBS Method

It has been pointed out already that (i) the fault-on trajectory (sustained trajectory) crosses the PEBS at a point close to the controlling u.e.p. and (ii) the potential energy $V_{PE}(\underline{\theta})$ of the post-fault system at this point is approximately equal to the value of the V-function at the controlling u.e.p. for the first swing instability. Hence, this value can be taken as a first approximation to V_{cr}. Therefore, the PEBS method of computing t_{cr} consists of the following steps.

(i) Compute the fault-on trajectory either by numerical integration or by one of the fast methods discussed in Secs. (5.4.1) and (5.4.4).

(ii) Compute $\underline{f}^T(\underline{\theta})\cdot(\underline{\theta}-\underline{\theta}^s)$ and $V_p(\underline{\theta}) + V_d(\underline{\theta}) = V_{PE}(\underline{\theta})$. $V_d(\underline{\theta})$ is computed using the approximations given by Eq. (5.54). Note that both in the computation of $V_{PE}(\underline{\theta})$ and $\underline{f}^T(\underline{\theta})\cdot(\underline{\theta}-\underline{\theta}^s)$, the parameters pertain to the post-fault configuration.

(iii) Inside PEBS, the expression $\underline{f}^T(\underline{\theta})\cdot(\underline{\theta}-\underline{\theta}^s)$ is less than zero and outside the PEBS it is greater than zero. Hence, steps (i) and (ii) are repeated until $\underline{f}^T(\underline{\theta})\cdot(\underline{\theta}-\underline{\theta}^s) = 0$. This is the PEBS crossing and the value of $V_{PE}(\underline{\theta})$ at this point is an estimate of V_{cr}.

(iv) With V_{cr} obtained in step (iii) integrate the faulted trajectory until $V(\underline{\tilde{\omega}},\underline{\theta}) = V_{cr}$ which yields t_{cr}.

Iterative PEBS Method

Following Ref. [2], a variation of the above method discussed below can improve the value of V_{cr}. At the end of step (iii), let the value of V_{cr} be designated as V_{cr1}. Then, integrate the faulted equations and clear the fault at $V(\underline{x}) = V_{cr1}$. Monitor the post-fault trajectory until the PEBS is reached. The value of $V_{PE}(\underline{\theta})$ at the PEBS crossing is the updated value of $V_{cr} = V_{cr2}$. Repeat this procedure two or three times to improve the estimate of V_{cr}. Two comments are pertinent here.

(i) Both the PEBS and the iterative PEBS methods are strictly valid for the first swing instability case. However, fairly accurate results are also obtained for multi-swing cases also.

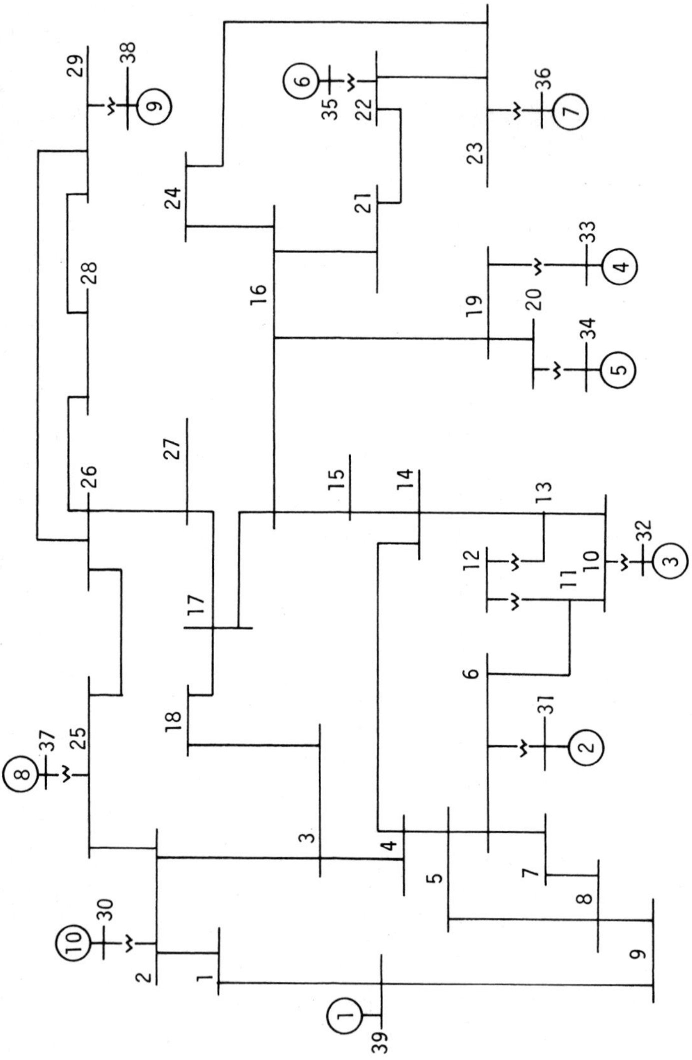

Fig. 5.9. 10 unit 39 bus New England test system.

(ii) In Ref. [2], instead of using the potential energy approximation for $V_d(\underline{\theta})$, the Lyapunov function due to Gudaru [19] is used. However, a correction to $V(\underline{x})$ is then applied to obtain V_{cr} since $\dot{V}(\underline{x})$ is sign indefinite. Also the PEBS is identified by the zero crossing of $\dot{V}_k$ instead of $\underline{f}^T(\underline{\theta})\cdot(\underline{\theta}-\underline{\theta}^s)$.

Numerical Example [5]

A 10-machine 39-bus system known as the New England Test System is used to illustrate the computation of t_{cr}. The single line diagram is given in Fig. 5.9. The line data and generator data are contained in many references in the literature. Two methods are used to compute the critical energy V_{cr} and the critical clearing time t_{cr}.

(i) Controlling u.e.p. method (Sec. 5.4.4).
(ii) PEBS method (Sec. 5.4.5).

For both these methods, the results of t_{cr} are compared with the actual t_{cr}. The results are shown in Table 5.4. The critical energy computed in column 6 is based on ascertaining the mode of instability for a given fault by simulation and then computing the exact u.e.p. V_{cr} is then obtained by evaluating $V(\underline{x})$ at this u.e.p. Except for multiple swing cases with severe intermachine oscillations due to the post-fault network, this u.e.p. coincides with the u.e.p. found from the u.e.p. method as outlined in Sec. 5.4.4. From Table 5.4 it is seen that there is good agreement between the actual critical clearing time and the critical clearing time using the u.e.p. method as well as the PEBS method. The PEBS method has the added advantage of being computationally fast. In fact, it is four times as fast as the u.e.p. method and about twenty times as fast as the conventional repetitive simulation method. This enhances the possibility of using this method for complementing simulation methods for planning as well as for on-line transient security monitoring purposes. Plots of system energies V_k and V_{PE}, $F(\underline{\theta})$, $\underline{f}^T(\underline{\theta})\cdot(\underline{\theta}-\underline{\theta}^s)$ are given in Fig. 5.10 for the fault at bus 31 both for stable and unstable cases. Fig. 5.11 shows the swing curves. It may be observed that for the stable case, (i) the $V_{PE}(\underline{\theta})$ and $V_k(\tilde{\omega})$

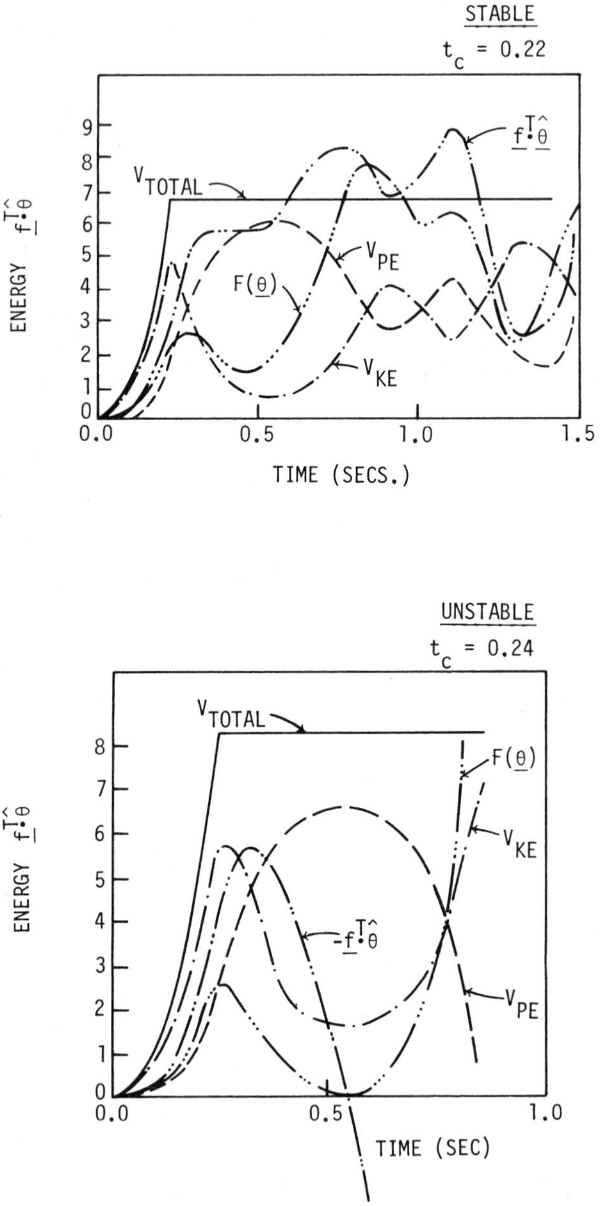

Fig. 5.10. Plots of system potential (PE), kinetic (KE), and total energy (V) and the quantities $F(\underline{\theta})$ and $\underline{f}^T(\underline{\theta}) \cdot (\underline{\theta} - \underline{\theta}^s)$ - classical system representation, case [31*], 10 unit system.

(Reproduced from Ref. [5]).

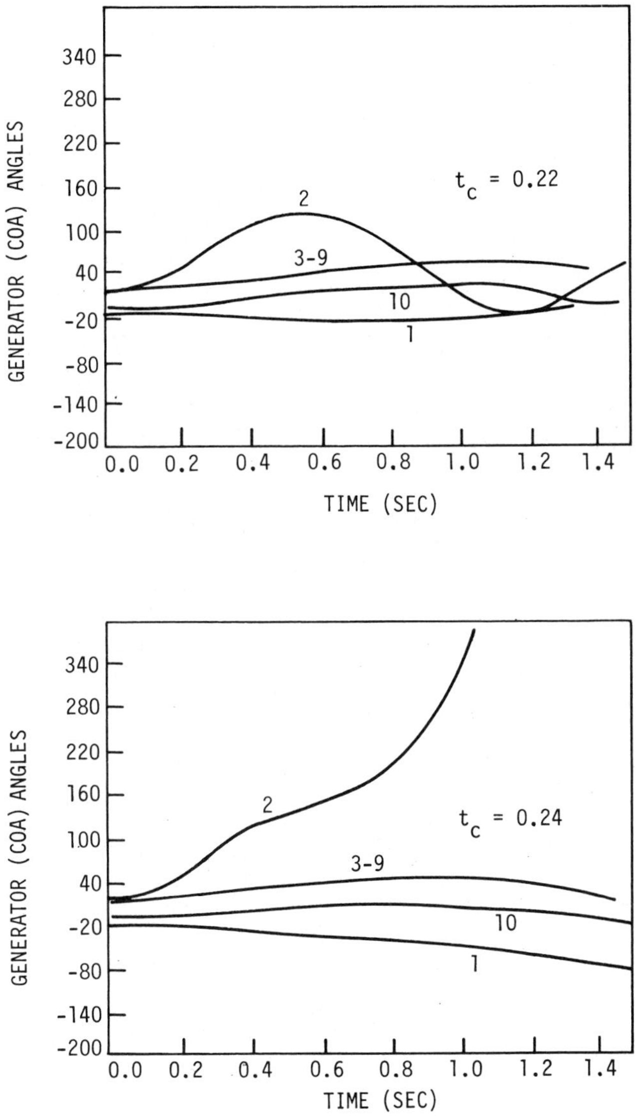

Fig. 5.11. Stable/unstable swing curves case [31*] - 10 unit system. (Reproduced from Ref. [5]).

reach maximum and minimum at the same time respectively, (ii) $\underline{F}(\underline{\theta})$ has an irregular pattern because the trajectory stays within the PEBS and (iii) $\underline{f}^T(\underline{\theta})\cdot(\underline{\theta}-\underline{\theta}^s)$ is always less than zero. For the critically unstable trajectory when $V_{PE}(\underline{\theta})$ reaches a maximum, V_k reaches a relative minimum and at the same instant of time, the mismatch $\underline{F}(\underline{\theta})$ in zero and $\underline{f}^T(\underline{\theta})\cdot(\underline{\theta}-\underline{\theta}^s)$ changes sign. Hence, the nature of PEBS and its properties relative to loss of synchronism are confirmed. The value of V_{PE} at the PEBS is about 6.5 which approximates well the value of V_{cr}.

Line Tripped *Faulted Bus	Critical Clearing time (secs)				Critical Energy	
	Simulation Stable	Simulation Unstable	Controlling UEP	PEBS	Controlling UEP	PEBS
2^*-3^m	0.24	0.26	0.235	0.26	8.99	12.7
4^*-14^m	0.22	0.24	0.215	0.225	11.75	10.1
6^*-11	0.20	0.22	0.21	0.195	9.31	7.95
15^*-16	0.22	0.24	0.23	0.225	10.47	10.1
$23-24^{*m}$	0.18	0.20	0.175	0.185	10.68	7.9
25^*-26^m	0.18	0.20	0.19	0.21	8.54	11.5
31	0.22	0.24	0.22^+(0.15)	0.21	6.77	6.35
35^m	0.24	0.25	0.25^+(0.14)	0.225	12.55	10.1
37^m	0.22	0.24	0.23^+(0.14)	0.215	10.7	9.7

m multi-swing case

\+ figures in brackets are by minimum energy u.e.p. method

Table 5.4 Critical Clearing Times for Faults in the 10-Machine, 39-Bus System

Conclusion

In this chapter, we have discussed the different methods of computing the region of attraction. We mainly considered those methods which took into account the faulted dynamics of the system in computing the critical energy. Methods prior to 1978 which did not consider the fault location have been briefly reviewed and their conceptual linkage established with the more recent methods. The two numerical examples illustrate the fact that Lyapunov function method is indeed a reliable tool in predicting accurately the critical clearing time for first swing classical model stability analysis.

References

1. El-Abiad, A. H. and Nagappan, K., "Transient Stability Regions for Multi-machine Power Systems", IEEE Transactions, Vol. PAS-85, 2, Feb. 1966, pp. 169-179.

2. Kakimoto, N., Ohsawa, Y and Hayashi, M., "Transient Stability Analysis of Electric Power Systems via Luré Type Lyapunov Function, Parts I and II", Trans. IEE of Japan, Vol. 98, No. 5/6, May/June 1978.

3. Athay, T., Podmore, R. and Virmani, S., "A Practical Method for the Direct Analysis of Transient Stability", IEEE Transactions, Vol. PAS-98, 2, March/April 1979, pp. 573-584.

4. Ribbens-Pavella, M., Grujic, Lj T., Sabatel, J. and Bouffioux, A., "Direct Methods for Stability Analysis of Large Scale Power Systems", Proceedings IFAC Symposium on 'Computer Applications in Large Scale Power Systems', New Delhi, India, Aug. 1979. Pergamon Press, Oct. 1980.

5. Athay, T., Sherkey, V. R., Podmore, R., Virmani, S. and Puech, C., "Transient Energy Stability Analysis", Conference on 'System Engineering for Power: Emergency Operating State Control - Section IV', Davos, Switzerland 1979, also U.S. Dept. of Energy Publication No. CONF-790904-PL 1980.

6. Willems, J. L., "Stability Theory of Dynamical Systems", (Book), Thomas Nelson & Sons Ltd., U.K. 1970.

7. Walker, J. A. and McLamroch, N. H., "Finite Regions of Attraction for Problem of Luré", Int. Journal of Control, Vol. 6, Oct. 1967, pp. 331-336.

8. Pai, M. A. and Narayana, C. L., "Finite Regions of Attraction for Multi-nonlinear Systems and its Application to the Power Systems Stability Problem", IEEE Trans. on Automatic Control, Oct. 1976, pp. 716-721.

9. Ribbens-Pavella, M., "On-line Measurements of Transient Stability Power System Index", in Savulescu S.C. (Editor) "Computerized Operation of Power Systems" (Book), Elsevier Scientific Publishing Co., New York 1976.

10. Uyemura, K. and Matsuki, J., "A Computational Algorithm for Evaluating Unstable Equilibrium States in Power Systems", Electrical Engineering in Japan, Vol. 92, No. 4, 1970, pp. 41-47.

11. Prabhakara, F. S. and El-Abiad, A. H., "A Simplified Determination of Transient Stability Regions for Lyapunov Methods", IEEE Trans., PAS Vol. PAS-94, Nc. 2, March/April 1975, pp. 672-680.

12. Gupta, C. L. and El-Abiad, A. H., "Determination of the Closest Equilibrium State for Lyapunov Methods in Transient Stability Studies", IEEE Trans., Vol. PAS-94, Sept./Oct. 1976, pp. 1699-1712.

13. Ribbens-Pavella, M. and Lemal, B., "Fast Determination of Stability Regions for On-line Power Systems Studies", Proc. IEE, Vol. 123, No. 7, July 1976, pp. 689-696.

14. Ribbens-Pavella, M., Howard, J. L., Lemal, B. and Nguyen, T. Ph., "Transient Stability Analysis by Scalar Lyapunov Functions: Recent Improvements and Practical Results", University of Liege, Faculty of Applied Science, Report No. 67-1977.

15. Doraiswami, R. and Fonseca, L.G.S., "A Fast and Reliable Dominion of Transient Stability for Multi-machine Power Systems", Paper A-77-060-7 IEEE Winter PES Power Meeting, New York, Feb. 1977.

16. Tavora, C. J. and Smith, O.J.M., "Stability Analysis of Power Systems", IEEE Trans., Vol. PAS-91, May/June 1972, pp. 1138-1145.

17. Varwandkar, S. D. and Pai, M. A., "Computational Aspects in Practical Application of Lyapunov Method in Power Systems", Proc. IFAC Symposium on 'Computer Applications in Large Scale Power Systems', New Delhi, Aug. 16-18, 1979, Pergamon Press 1980.

18. Tavora, C. J. and Smith, O.J.M., "Equilibrium Analysis of Power Systems, IEEE Trans., Vol. PAS-91, May/June 1971, pp. 1131-1137.

19. Gudaru, U., "A General Lyapunov Function for Multi-machine Power Systems with Transfer Conductances", Int. Journal of Control, Vol. 21, No. 2, Feb. 1975, pp. 333-343.

Chapter VI

STABILITY OF LARGE SCALE POWER SYSTEMS BY DECOMPOSITION

6.1 Introduction

The rapidly increasing size of power systems and the complex modes of instabilities that are inherent in such systems has emphasized the need to approach the stability analysis problem in a hierarchial manner. Such an approach involves decomposition of the composite system into subsystems which consist of lower order, isolated (free) subsystems plus the interconnections between the subsystems. The decomposition can be performed mathematically or physically. A Lyapunov function is constructed for each of the free subsystems which becomes much easier. After this there are two approaches possible to ascertain stability of the overall composite system. In the first approach, a scalar Lyapunov function consisting of the weighted sum of the free subsystem Lyapunov functions is constructed and its negative definiteness is examined along trajectories of the composite system. The interconnections influence the negative definiteness of the weighted sum scalar Lyapunov functions (WSSLF). In the second approach, a vector Lyapunov function (VLF) whose components are the subsystem Lyapunov functions is postulated. The derivative of the vector Lyapunov function is examined and applying the comparison principle conditions are obtained for the stability of the composite system. Under certain mathematical assumptions, it can be shown that both the approaches yield equivalent results.

There has been a vast literature on the subject of stability of large scale dynamical systems. The concept of vector Lyapunov functions originally proposed by Bellman [1] was first applied to interconnected dynamical systems by

Bailey [2]. Since then, for nearly a decade there has been extensive literature on this topic which has been discussed authoritatively in two books, Michel and Miller [3] and Siljak [4]. In this chapter, our focus will be on the application of both the weighted sum scalar Lyapunov function and the vector Lyapunov function approaches to the power system stability problem. Both of these methods share a common principle, namely, the decomposition of the composite system into subsystems. In that context, we shall first study methods of decomposing a power system. These range from purely mathematical ones to those based on physical insight such as coherency, weak tie lines, etc. Next, we consider application of both the weighted sum scalar Lyapunov function and the comparison principle based vector Lyapunov function.

We have seen in Chapter V that estimating the region of stability is a very important step in computing the critical clearing time. In decomposition based methods, it is attractive to compute the region of stability of the composite system from regions of stability of the individual free subsystems. We use results of Weissenberger [5] in this connection.

6.2 Decomposition Techniques

Application of the vector Lyapunov function approach to the transient stability problem of a power system was first initiated by Pai and Narayana [6]. In their approach, a n-machine system was decomposed into n(n-1)/2 subsystems. Thus, for higher order systems, the most important advantage of the vector Lyapunov function, namely, reducing the dimensionality of the stability problem is thus lost. However, as we shall see later, their decomposition was essentially an overdescription of the problem. Later on, a number of other decompositions were proposed which greatly reduced the number of systems. References [7,8,9,10] propose a number of variations of decompositions into (n-1) or n subsystems depending on whether the system has uniform or non-uniform damping. Reference [10] includes governor action also in the subsystems.

These decomposition procedures can take into account transfer conductances in their models. If we ignore transfer conductances, then other types of decomposition such as those preceded by an equivalencing procedure to group machines into coherent groups are possible [11,12]. Since the primary advantage of the stability analysis via decomposition is the possibility of modelling subsystems more accurately, we restrict our treatment to systems which do not simplify the composite model to start with.

6.2.1 Decomposition with Non-uniform Damping

Consider a n-machine system whose swing equation is given in Eq. (3.13) and is reproduced below

$$M_i \frac{d^2\delta_i}{dt^2} + D_i \frac{d\delta_i}{dt} = P_i - P_{ei}$$

where $P_i = P_{mi} - |E_i|^2 G_{ii}$

$$P_{ei} = \sum_{\substack{j=1\\ \neq i}}^{n} |E_i||E_j||Y_{ij}|\cos(\delta_{ij}-\phi_{ij}) \tag{6.1}$$

$$= \sum_{\substack{j=1\\ \neq i}}^{n} C_{ij}\sin\delta_{ij} + D_{ij}\cos\delta_{ij}$$

If the parameters in Eq. (6.1) correspond to the post-fault state, then the post-fault SEP δ^s can be transferred to the origin and the Eq. (6.1) can be rewritten as Eq. (3.35), i.e.

$$M_i \frac{d^2\delta_i}{dt^2} + D_i \frac{d\delta_i}{dt} = P_{ei}(\underline{\delta}^s) - P_{ei}(\underline{\delta}), \quad i=1,2,\ldots,n \tag{6.2}$$

$$= \sum_{\substack{j=1\\ \neq i}}^{n} - A_{ij}[\cos(\delta_{ij}-\phi_{ij}) - \cos(\delta_{ij}^s-\phi_{ij})] = \sum_{\substack{j=1\\ \neq i}}^{n} - A_{ij}f_{ij}$$

where $A_{ij} \triangleq |E_i||E_j||Y_{ij}|$

$f_{ij} \triangleq \cos(\delta_{ij}-\phi_{ij}) - \cos(\delta_{ij}^s-\phi_{ij})$

A_{ij}, C_{ij} and D_{ij} are related as follows:

$$C_{ij} = |E_i||E_j|B_{ij} = |E_i||E_j||Y_{ij}|\sin\phi_{ij} = A_{ij}\sin\phi_{ij}$$

$$D_{ij} = |E_i||E_j|G_{ij} = |E_i||E_j||Y_{ij}|\cos\phi_{ij} = A_{ij}\cos\phi_{ij} \tag{6.3}$$

To facilitate the understanding of later material, we write Eq. (6.2) corresponding to the nth machine as

$$M_n \frac{d^2\delta_n}{dt^2} + D_n \frac{d\delta_n}{dt} = P_{en}(\underline{\delta}^s) - P_{en}(\underline{\delta})$$

$$= -\sum_{j=1}^{n-1} A_{nj}f_{nj} \tag{6.4}$$

$$= -\sum_{j=1}^{n-1} A_{nj}[\cos(\delta_{nj}-\phi_{nj}) - \cos(\delta_{nj}^s-\phi_{nj})]$$

Define $\frac{D_i}{M_i} = \lambda_i$ and the state variables as δ_{in}, ω_{in} $(i=1,2,\ldots,n-1)$ and ω_n. The (2n-1) dimensional state space model can be obtained as

$$\dot{\delta}_{in} = \omega_n$$

$$\dot{\omega}_{in} = -\lambda_i\omega_{in} + (\lambda_n-\lambda_i)\omega_n$$

$$+ M_n^{-1} \sum_{j=1}^{n-1} A_{nj}f_{nj} - M_i^{-1} \sum_{\substack{j=1 \\ \neq i}}^{n} A_{ij}f_{ij}$$

$$\dot{\omega}_{in} = -\lambda_n\omega_n - M_n^{-1} \sum_{j=1}^{n-1} A_{nj}f_{nj} \qquad i=1,2,\ldots,n-1 \tag{6.5}$$

The most effective decomposition which takes into account the transfer conductances in the model is the pair-wise decomposition [7-10]. There are two possibilities.

(i) The choice of subsystem state vector is $\underline{x}_i = (\delta_{in}-\delta^s_{in}, \omega_{in}, \omega_n)$, i=1,2,...,n-1. We impose the restriction on the choice of the reference machine that $\lambda_n \geq \lambda_i$ (i=1,2,...,n-1). With this choice of subsystem state vector, we have (n-1) subsystems. The state variable ω_n corresponding to the reference machine appears in all the subsystem descriptions. In the literature [8] this is often called the overlapping decomposition.

(ii) The overlapping decomposition can be avoided as indicated in Ref. [10] if we define n subsystems such that the state vector for the first (n-1) subsystems is $\underline{x}_i = (\delta_{in}-\delta^s_{in}, \omega_n)$ and the state vector for the nth subsystem is the scalar ω_n.

We first consider decomposition of the type (i).

In (6.5), the term corresponding to $M_n^{-1}A_{ni}f_{ni}$ and $M_i^{-1}A_{in}f_{in}$ be expressed as

$$\begin{aligned} M_n^{-1}A_{ni}f_{ni} &= M_n^{-1}|E_i||E_n||Y_{in}| \\ &\quad [\cos(\delta_{ni}-\phi_{ni}) - \cos(\delta^s_{ni}-\phi^s_{ni})] \\ &= -M_n^{-1}C_{in}(\sin\delta_{in}-\sin\delta^s_{in}) \\ &\quad + M_n^{-1}D_{in}(\cos\delta_{in}-\cos\delta^s_{in}) \end{aligned} \tag{6.6}$$

and

$$\begin{aligned} M_i^{-1}A_{in}f_{in} &= M_i^{-1}C_{in}(\sin\delta_{in}-\sin\delta^s_{in}) \\ &\quad + M_i^{-1}D_{in}(\cos\delta_{in}-\cos\delta^s_{in}) \end{aligned} \tag{6.7}$$

We now have the ith subsystem description with $\underline{x}_i^T = (\delta_{in}-\delta^s_{in}, \omega_{in}, \omega_n)$ as

$$\dot{\underline{x}}_i = \begin{bmatrix} 0 & 1 & 0 \\ 0 & -\lambda_i & \lambda_n-\lambda_i \\ 0 & 0 & -\lambda_n \end{bmatrix} \underline{x}_i$$

$$+ \begin{bmatrix} 0 \\ -(M_n^{-1}+M_i^{-1})C_{in} \\ M_n^{-1}C_{in} \end{bmatrix} \underline{f}_i(\sigma_i) + \underline{h}_i(\underline{x}) \qquad (6.8)$$

$$\sigma_i = [1 \quad 0 \quad 0]\underline{x}_i \qquad i=1,2,\ldots,n-1$$

where

$$f_i(\sigma_i) = \sin(\sigma_i+\delta_{in}^s) - \sin\delta_{in}^s \qquad (6.9)$$

$$\underline{h}_i(\underline{x}) = \begin{bmatrix} 0 \\ (M_n^{-1}-M_i^{-1})D_{in}\phi_i + \sum\limits_{\substack{j=1 \\ \neq i}}^{n-1} (M_n^{-1}A_{nj}f_{nj}-M_i^{-1}A_{ij}f_{ij}) \\ -M_n^{-1}D_{in}\phi_i - \sum\limits_{\substack{j=1 \\ \neq i}}^{n-1} M_n^{-1}A_{nj}f_{nj} \end{bmatrix} \qquad (6.10)$$

$$\phi_i = (\cos(\sigma_i+\delta_{in}^s)-\cos\delta_{in}^s) \qquad (6.11)$$

In the description (6.8), the last equation is common to all the subsystems. This can be avoided by defining the subsystem state vectors as $\underline{x}_i^T = (\delta_{in}-\delta_{in}^s,\omega_{in})$ $i=1,2,\ldots,n-1$ and $x_n = \omega_n$. Now we write the ith subsystem description $(i=1,2,\ldots,n-1)$ as

$$\dot{\underline{x}}_i = \begin{bmatrix} 0 & 1 \\ 0 & -\lambda_i \end{bmatrix} \underline{x}_i$$

$$+ \begin{bmatrix} 0 \\ -(M_n^{-1}+M_i^{-1})C_{in} \end{bmatrix} f_i(\sigma_i) + [\underline{h}_i(\underline{x})]$$

$$+ \begin{bmatrix} 0 \\ (\lambda_n-\lambda_i)x_n \end{bmatrix}$$

$$\dot{x}_n = -\lambda_n x_n - M_n^{-1} \sum_{j=1}^{n-1} A_{nj} f_{nj} \qquad i=1,2,\ldots,n-1 \qquad (6.12)$$

The difference between this description and (6.8) is easily seen by inspection.

We have only second order subsystems to deal with as opposed to third order subsystems. A more general form of this decomposition with asynchronous damping and governor representation is contained in Refs. [9,10].

6.2.2 Decomposition with Uniform Damping [7]

In this case, $\lambda_i = \lambda$ for all i. The state vector for the composite system is $\underline{x}^T = (\delta_{in}-\delta_{in}^s,\ldots,\delta_{n-1,n}-\delta_{n-1,n}^s,$ $\omega_{in},\ldots,\omega_{n-1,n})$. The state space model is given by

$$\dot{\delta}_{in} = \omega_{in}$$

$$\dot{\omega}_{in} = -\lambda\omega_{in} + M_n^{-1} \sum_{j=1}^{n-1} A_{nj} f_{nj}$$

$$- M_i^{-1} \sum_{\substack{j=1 \\ \neq i}}^{n} A_{ij} f_{ij} \qquad i=1,2,\ldots,n-1 \qquad (6.13)$$

The subsystem vector is $\underline{x}_i = (\delta_{in}-\delta_{in}^s,\omega_{in})$ and the subsystem state space model becomes

$$\dot{\underline{x}}_i = \begin{bmatrix} 0 & 1 \\ 0 & -\lambda \end{bmatrix} \underline{x}_i + \begin{bmatrix} 0 \\ -1 \end{bmatrix} \hat{f}_i(\sigma_i) + \hat{\underline{h}}_i(\underline{x}) \tag{6.14}$$

$$\sigma_i = [1 \quad 0]\underline{x}_i \qquad i=1,2,\ldots,n-1$$

$$f_i(\sigma_i) = (M_n^{-1}+M_i^{-1})C_{in}[\sin(\sigma_i+\delta_{in}^s) - \sin\delta_{in}^s] \tag{6.15}$$

$$\hat{\underline{h}}_i(\underline{x}) = \begin{bmatrix} 0 \\ (M_n^{-1}-M_i^{-1})D_{in}\phi_i + \sum\limits_{\substack{j=1 \\ \neq i}}^{n-1} (M_n^{-1}A_{nj}f_{nj}-M_i^{-1}f_{ij}A_{ij}) \end{bmatrix} \tag{6.16}$$

where ϕ_i is defined as in (6.11). Equation (6.14) is the simplest decomposition where the free subsystems do not contain any transfer conductance terms. Two improvements over this decomposition are possible.

(i) Absorbing transfer conductances into the free subsystem [13]

We rewrite (6.14) as

$$\dot{\underline{x}}_i = \begin{bmatrix} 0 & 1 \\ 0 & -\lambda \end{bmatrix} \underline{x}_i + \begin{bmatrix} 0 \\ -1 \end{bmatrix} f_i(\sigma_i) + \underline{h}_i(\underline{x}) \tag{6.17}$$

$$\sigma_i = [1 \quad 0]\underline{x}_i$$

where

$$f_i(\sigma_i) = (M_n^{-1}+M_i^{-1})C_{in}[\sin(\sigma_i+\delta_{in}^s) - \sin\delta_{in}^s] + (M_i^{-1}-M_n^{-1})D_{in}[\cos(\sigma_i+\delta_{in}^s) - \cos\delta_{in}^s] \tag{6.18}$$

$$\underline{h}_i(\underline{x}) = \begin{bmatrix} 0 \\ \sum\limits_{\substack{j=1 \\ \neq i}}^{n-1} (M_n^{-1}A_{nj}f_{nj}-M_i^{-1}A_{ij}f_{ij}) \end{bmatrix}$$

Denoting $(M_n^{-1}+M_i^{-1})C_{in} = \mu_{1i}$ and $(M_i^{-1}-M_n^{-1})D_{in} = \mu_{2i}$, (6.18) can be rewritten as

$$f_i(\sigma_i) = \sqrt{\mu_{1i}^2 + \mu_{2i}^2}\ \{\sin(\sigma_i+\delta_{in}^s+\psi_{in}) - \sin(\delta_{in}^s+\psi_{in})\} \tag{6.19}$$

where $\tan\psi_{in} = (M_i^{-1}-M_n^{-1})D_{in}/(M_n^{-1}+M_i^{-1})C_{in}$

If $|\delta_{in}^s + \psi_{in}| < \pi/2$, then $f_i(\sigma_i)$ is in the first and third quadrant in an interval around the origin. If M_n is chosen as the machine with the largest inertia, the above condition is likely to be satisfied.

(ii) Modification of (6.18)

In this subsystem description, we delete the terms corresponding to M_n^{-1} in $f_i(\sigma_i)$ in (6.18) and include it in the interconnection term $h_i(\underline{x})$. Thus with (6.17) remaining the same, $f_i(\sigma_i)$ in (6.18) becomes

$$\begin{aligned} f_i(\sigma_i) &= M_i^{-1}\{C_{in}[\sin(\sigma_i+\delta_{in}^s) - \sin\delta_{in}^s] \\ &\quad + D_{in}[\cos(\sigma_i+\delta_{in}^s) - \cos\delta_{in}^s]\} \qquad (6.20) \\ &= M_i^{-1}A_{in}\{\sin(\sigma_i+\delta_{in}^s+\psi_{in}) - \sin(\delta_{in}^s+\psi_{in})\} \end{aligned}$$

where $\psi_{in} = \tan^{-1}\dfrac{D_{in}}{C_{in}} = \dfrac{\pi}{2} - \phi_{in}$

ϕ_{in} being the angle of the transfer conductance between internal buses i and n. Under these conditions $\underline{h}_i(\underline{x})$ becomes

$$\begin{aligned} \underline{h}_i(\underline{x}) &= [-M_n^{-1}A_{in}\{\sin(\sigma_i+\delta_{in}^s+\psi_{in}) + \sin(\delta_{in}^s+\psi_{in})\} \\ &\quad + \sum_{\substack{j=1\\ \neq i}}^{n-1} (M_n^{-1}A_{nj}f_{nj}-M_i^{-1}A_{ij}f_{ij})] \end{aligned} \tag{6.21}$$

We thus see that there are quite a few variations in constructing the free subsystem structure in pair-wise decomposition techniques.

6.2.3 Other Decomposition Techniques

(i) In a pair-wise decomposition structure, we can in addition take into account asynchronous damping and first order proportional governors [10].

(ii) One of the restrictions in the non-uniform damping case is that $\lambda_n \geq \lambda_i$. This is necessary in order that the linear part of the free subsystem in (6.8) be stable. This restriction can be removed if instead of ω_n we introduce ω_o which is the velocity corresponding to the center of angle. A new set of decomposition and its variations is possible [14].

(iii) If we do an apriori coherency analysis according to some criteria and group the machines, then the number of subsystems could be reduced. Lyapunov function can be applied to the reduced model. This technique discussed in Refs. [11] and [12] however does not strictly fall into the scope of our discussion.

(iv) All the previous techniques rely on a mathematical decomposition of the composite system. Since the description is based on the bus admittance matrix at the internal nodes after eliminating all physical nodes, any type of physical decomposition is therefore ruled out. However, if a structure preserving model is used as in Ref. [15], then a physical decomposition is possible [16].

Although a great number of decomposition techniques have been discussed in the literature, there is no clear cut evidence to indicate that any one method is preferable. Since our main focus is on methods which have met with some success in practical application, detailed discussion of decomposition techniques has been limited to such cases only.

6.3 Weighted Sum Scalar Lyapunov Function Approach

We first state some general results before applying it for the power system problem. The treatment in this section follows largely that in Michel and Miller [3].

Consider the composite system

$$\dot{\underline{x}} = \underline{f}(\underline{x}) \qquad (6.22)$$

where $\underline{x}$ and $\underline{f}$ are N-vectors. Let the system (6.22) be decomposed into k subsystems with $\underline{x}_i$ of dimension n_i as the state vector of the ith subsystem. Hence,

$$\sum_{i=1}^{k} n_i = N$$

and $\underline{x} = (\underline{x}_1^T, \underline{x}_2^T, \ldots, \underline{x}_k^T)^T.$

Let the decomposition have the form

$$\dot{\underline{x}}_i = \underline{f}_i(\underline{x}_i) + \underline{g}_i(\underline{x}) \qquad i=1,2,\ldots,k \qquad (6.23)$$

where $\underline{g}_i(\underline{x})$ represents the interconnection term in the ith subsystem. A particular form of $\underline{g}_i(\underline{x})$ is

$$\sum_{\substack{j=1\\ \neq i}}^{k} g_{ij}(\underline{x})$$

$$\dot{\underline{x}}_i = \underline{f}_i(\underline{x}_i) \qquad (6.24)$$

represents the ith free subsystem.

The following theorem due to Michel [17] is fundamental to the weighted sum scalar Lyapunov function approach.

Theorem 6.1

The equilibrium $\underline{x} = \underline{0}$ of the composite system (6.23) is a.s.i.l. if the following conditions are satisfied.

(i) Each free subsystem (6.24) is asymptotically stable in the large with a Lyapunov function $v_i(\underline{x}_i) > 0$ and

$$\dot{v}(\underline{x}_i)\Big|_{(6.24)} \leq -\sigma_i \|\underline{x}_i\|^2 \quad (\sigma_i > 0) \text{ for all } \underline{x}_i \qquad (6.25)$$

where $\|\cdot\|$ denotes any one of the equivalent norms in R^{n_i}.

(ii) Given $v_i(\underline{x}_i)$, there exists constants a_{ij} such that the following inequality is satisfied with respect to the interconnection terms

$$\underline{\nabla V}(\underline{x}_i)^T \underline{g}_i(\underline{x}) \leq ||\underline{x}_i|| \sum_{j=1}^{k} a_{ij} ||\underline{x}_j|| \tag{6.26}$$

(iii) Given σ_i of hypothesis (i) there exists a k-vector $\underline{\alpha}^T = (\alpha_1,\ldots,\alpha_k) > 0$ such that the test matrix $\underline{S} = [s_{ij}]$ specified by

$$s_{ij} = \begin{cases} \alpha_i(-\sigma_i + a_{ii}) & i=j \\ (\alpha_i a_{ij} + \alpha_j a_{ji})/2 & i \neq j \end{cases} \tag{6.27}$$

is negative definite.

<u>Proof:</u> For the composite system, choose the Lyapunov function

$$V(\underline{x}) = \sum_{i=1}^{k} \alpha_i v_i(\underline{x}_i) \tag{6.28}$$

Clearly $V(\underline{x}) > 0$ in the entire region by virtue of hypothesis (i) and the fact that $\alpha_i > 0$ $(i=1,\ldots,k)$. Along the solutions of (6.23) we have

$$\begin{aligned} \dot{V}(x) &= \sum_{i=1}^{k} \alpha_i \{\underline{\nabla V}_i(\underline{x}_i)^T [\underline{f}_i(\underline{x}_i) + \underline{g}_i(\underline{x})]\} \\ &\leq \sum_{i=1}^{k} \alpha_i \{-\sigma_i ||\underline{x}_i||^2 + ||\underline{x}_i|| \sum_{j=1}^{k} a_{ij} ||\underline{x}_j||\} \end{aligned} \tag{6.29}$$

Now let $\underline{w}^T = (||\underline{x}_1||,\ldots,||\underline{x}_k||)$ and $\underline{R} = [r_{ij}]$ be the k×k matrix specified by

$$r_{ij} = \begin{cases} \alpha_i(-\sigma_i + a_{ii}) & i=j \\ \alpha_i a_{ij} & i \neq j \end{cases} \tag{6.30}$$

We now have

$$\dot{V}(\underline{x}) \leq \underline{w}^T \underline{R}\, \underline{w} = \underline{w}^T \left(\frac{\underline{R}+\underline{R}^T}{2}\right) \underline{w} = \underline{w}^T \underline{S}\, \underline{w} \leq \lambda_M(\underline{S})\, ||\underline{w}||^2 \tag{6.31}$$

where $\underline{S}$ is the test matrix of hypothesis (iii) and λ_M denotes the maximum eigenvalue. Hence $\dot{V}(\underline{x})$ is negative definite if $\underline{S}$ is negative definite. We also note that since $\underline{S}$ is symmetric, all its eigenvalues are real. Since $\underline{S}$ is negative definite, $\lambda_M(\underline{S}) < 0$.

The condition for negative definiteness of $\underline{S}$ is given by the following inequalities

$$(-1)^{\ell} \begin{vmatrix} s_{11} & \cdots & s_{1\ell} \\ \\ s_{\ell 1} & \cdots & s_{\ell\ell} \end{vmatrix} > 0 \qquad \ell=1,2,\ldots,k \tag{6.32}$$

6.4 Region of Attraction

Suppose that the hypothesis (i) of Theorem 6.1 is not valid in the entire state space. In other words, the origin of the ith free subsystem is asymptotically stable in a region around the state space and the region of attraction is defined as $v_i(\underline{x}_i) < v_{oi}$. Then, as shown by Weissenberger [5], the region of attraction of the composite system is given by

$$V(\underline{x}) < \gamma \quad \text{where} \quad \gamma = \min_{1 \le i \le k} (\alpha_i v_{oi}) \tag{6.33}$$

This region of attraction is in general conservative and one way of improving it is to optimize the α_i's, a topic to be discussed later.

6.5 Results Involving M-Matrices

It may be observed that in the application of Theorem 6.1, we need the existence of suitable α_i's > 0 (i=1,2,...,k). In the special case when the off diagonal elements of the test matrix $\underline{S}$ are ≥ 0 (non-negative) we can utilize properties of the so-called M-matrices [3] to establish results which are easier to apply because (i) they do not involve usage of weightage vector $\underline{\alpha}$ and (ii) under certain conditions the results are equivalent to those obtained by the comparison principle and vector Lyapunov function. We first review certain properties of M-matrices.

M-Matrices [3]

Definition

A real k×k matrix $\underline{D} = [d_{ij}]$ is said to be an M-matrix if (i) $d_{ij} \leq 0$ $i \neq j$ (i.e. all off diagonal elements of $\underline{D}$ are non-positive) and (ii) if all principal minors are positive, i.e. if D_ℓ denotes the determinant

$$D_\ell = \begin{vmatrix} d_{11} & \cdots & d_{1\ell} \\ & & \\ d_{\ell 1} & \cdots & d_{\ell 1} \end{vmatrix} > 0 \qquad \ell = 1,2,\ldots,k \qquad (6.34)$$

The following equivalent statements bring out some of the properties of M-matrices.

a) $\underline{D}$ is an M-matrix.
b) There is a k-vector $\underline{u} > 0$ such that $\underline{D}\underline{u} > 0$.
c) There is a k-vector $\underline{v} > 0$ such that $\underline{D}^T\underline{v} > 0$.
d) The real parts of all eigenvalues of $\underline{D}$ are positive.
e) $\underline{D}$ is nonsingular and all elements of $\underline{D}^{-1}$ are non-negative (in fact, all diagonal elements of $\underline{D}^{-1}$ are positive).

From (b) and (c) it also follows, respectively that

(i) There exist positive constants λ_j (j=1,...,k) such that

$$\sum_{j=1}^{k} \lambda_j d_{ij} > 0 \qquad i=1,2,\ldots,k$$

(ii) There exist positive constants η_j (j=1,2,...,k) such that

$$\sum_{j=1}^{k} \eta_j d_{ji} > 0 \qquad i=1,2,\ldots,k$$

Another useful property of M-matrices which we require is the following.

Corollary. Let $\underline{D} = [d_{ij}]$ be a k×k matrix with non-positive off diagonal elements, i.e. $d_{ij} \leq 0$ $i \neq j$. Then $\underline{D}$ is an M-matrix if and only if there exists a diagonal matrix $\underline{A} = \text{diag}(\alpha_1,\ldots,\alpha_k)$, $\alpha_i > 0$ i=1,2,...,k such that the matrix

$$\underline{B} = \underline{D}^T\underline{A} + \underline{A}\,\underline{D} \qquad (6.35)$$

is positive definite.

This concludes the discussion on M-matrices and its properties. We now reformulate the stability Theorem 6.1 in terms of M-matrices following Michel and Miller [3].

Theorem 6.2 [3]

Suppose that conditions (i) and (ii) of Theorem 6.1 are true with the additional assumption that $a_{ij} \geq 0$ for all $i \neq j$. Then the equilibrium $\underline{x} = \underline{0}$ of (6.23) is a.s.i.l. if any one of the following conditions hold

(i) The successive principal minors of the $k \times k$ test matrix $\underline{D} = [d_{ij}]$ are all positive, where

$$d_{ij} = \begin{cases} \sigma_i - a_{ii} & i=j \\ \\ -a_{ij} & i \neq j \end{cases} \tag{6.36}$$

(ii) The real parts of the eigenvalues of $\underline{D}$ are all positive.

(iii) There exist positive constants λ_i $(i=1,\ldots,k)$ such that

$$(\sigma_i - a_{ii}) - \sum_{\substack{j=1\\ \neq i}}^{k} \frac{\lambda_j}{\lambda_i} a_{ij} > 0 \qquad i=1,2,\ldots,k \tag{6.37}$$

(iv) There exist positive constants η_i $(i=1,2,\ldots,k)$ such that

$$(\sigma_i - a_{ii}) - \sum_{\substack{j=1\\ \neq i}}^{k} \frac{\eta_j}{\eta_i} a_{ji} > 0 \qquad i=1,2,\ldots,k \tag{6.38}$$

Proof. Since by assumption $a_{ij} \geq 0$ for $i \neq j$ and since the successive principal minors of $\underline{D}$ are positive, it follows that $\underline{D}$ is an M-matrix. Hence, from the property of M-matrices, there exists a diagonal matrix $\underline{A} = \text{diag}(\alpha_1,\ldots,\alpha_k) > 0$ such that

$$-2\underline{S} = \underline{D}^T\underline{A} + \underline{A}\,\underline{D} \tag{6.39}$$

is positive definite.

Hence $-2\underline{S} = -2[s_{ij}]$ where

$$s_{ij} = \begin{cases} \alpha_i(-\sigma_i+a_{ii}) & i=j \\ (\alpha_i a_{ij}+\alpha_j a_{ji})/2 & i\neq j \end{cases} \tag{6.40}$$

is positive definite. If $-2\underline{S}$ is positive definite, $\underline{S}$ is negative definite. This tantamounts to satisfying condition (iii) of Theorem 6.1. Hence, the equilibrium is a.s.i.l. Conditions (ii) - (iv) are true since $\underline{D}$ is an M-matrix. Equations (6.37) and (6.38) are referred to as row dominance and column dominance conditions, respectively.

It may be noted that condition (i) is independent of α_i's. Also, (6.37) and (6.38) can be used to optimize the region of attraction as shown below.

6.6 Optimizing the Region of Attraction

The region of attraction for the composite system has been defined to be

$$V(\underline{x}) < \gamma \quad \text{where} \quad \gamma = \min_{1\leq i\leq k} (\alpha_i v_{oi})$$

To get the best estimate, we need to optimize α_i's. We may associate α_i's with either λ_i's or η_i's in (6.37) and (6.38). Suppose $\alpha_i = \lambda_i$, then since $\alpha_i > 0$ $(i=1,\ldots,k)$ and $a_{ij} \geq 0$ $(i\neq j)$ it follows from (6.37) that $\sigma_i - a_{ii} > 0$ if the origin is asymptotically stable. As suggested by Chen and Schinzinger [13] the row dominance property (6.37) (called also as "negative diagonal quasidominance" [4]) can be used to produce the following two criteria for optimizing the region of attraction.

Criterion 1

Minimize the trace of the test matrix $\underline{S}$

$$\min_{\underline{\alpha}} z = \sum_{i=1}^{k} \alpha_i(\sigma_i-a_{ii}) \tag{6.41}$$

subject to (6.37) namely

$$\alpha_i(\sigma_i - a_{ii}) - \sum_{\substack{j=1 \\ \neq i}}^{k} \alpha_j a_{ij} > 0 \qquad i=1,2,\ldots,k$$

$$\sum_{i=1}^{k} \alpha_i = 1 \tag{6.42}$$

Criterion 2

Maximize the weighted sum of the boundaries of the stability region of the subsystems

$$\operatorname*{Max}_{\underline{\alpha}} \; z = \sum_{i=1}^{k} \alpha_i v_{oi} \tag{6.43}$$

subject to (6.42).

Let α_i^* be the optimized value of α_i's by either criterion 1 or 2. The region of stability is defined to be

$$\sum_{i=1}^{k} \alpha_i^* v_i(\underline{x}_i) < V^\circ \tag{6.44}$$

and $$V^\circ = \operatorname*{Min}_{1 \le i \le k} (\alpha_i^* v_{oi})$$

In view of the numerous equivalent properties of M-matrices, there are quite a few other possibilities for optimizing the α_i's.

6.7 Application to Power Systems [13]

Since the best results to date have been obtained by the weighted sun scalar Lyapunov function method using decomposition, we present a numerical example in detail. We consider a uniformly damped system and first state the general formulation for an n-machine system before presenting the example. The presentation follows closely that of Chen and Schinzinger [13]. The subsystem modelling is the same as in (6.17) which involves absorbing the transfer conductances in the free subsystems.

$$\dot{\underline{x}}_i = \underline{A}_i \underline{x}_i + \underline{b}\, f_i(\sigma_i) + \underline{h}_i(x)$$

$$\underline{\sigma}_i = [1 \quad 0]\underline{x}_i = \underline{c}_i^T \underline{x}_i \tag{6.45}$$

$$\text{where} \quad \underline{x}_i = \begin{bmatrix} x_{1i} \\ x_{2i} \end{bmatrix} = \begin{bmatrix} \delta_{in} - \delta_{in}^s \\ \omega_{in} \end{bmatrix}; \quad \underline{A}_i = \begin{bmatrix} 0 & 1 \\ 0 & -\lambda \end{bmatrix}$$

$$\underline{b}^T = [0 \quad -1] \qquad i=1,2,\ldots,n-1$$

$\underline{f}_i(\underline{\sigma}_i)$ and $\underline{h}_i(\underline{x})$ are defined in (6.18) and (6.19). The machine having the largest inertia is designated as the nth machine and chosen as the reference machine. Hence $M_n^{-1} << M_i^{-1}$ so that ψ_{in} in (6.19) is almost equal to $\frac{\pi}{2} - \theta_{in}$ where θ_{in} is the transfer admittance angle between machine i and n.

6.7.1 Subsystem Analysis

The ith free subsystem is described by

$$\dot{\underline{x}}_i = \underline{A}_i\underline{x}_i + \underline{b}\, f_i(\sigma_i)$$
$$\sigma_i = \underline{c}_i^T\underline{x}_i \qquad i=1,2,\ldots n-1 \tag{6.46}$$

$f_i(\sigma_i)$ lies in the first and third quadrants in a region around the origin defined by $-\pi - 2(\delta_{in}^s+\psi_{in}) < \sigma_i < \pi - 2(\delta_{in}^s+\psi_{in})\cdot v_i(\underline{x}_i)$ is constructed following the procedure in Sec. 4.6 since Eq. (6.46) is analogous to Eq. (4.37) with the following equivalences

$$-D/M = \lambda; \quad M = \frac{1}{\sqrt{\mu_{1i}^2 + \mu_{2i}^2}} = M_{eq}; \quad \delta^s = \delta_{in}^s + \psi_{in}^s \tag{6.47}$$

The most general form of the Lyapunof function is used, namely Eq. (4.44). Choosing n=2 and dividing by M, we obtain from Eq. (4.44) the Lyapunov function for the system (6.44) as

$$v_i(\underline{x}_i) = \frac{1}{2}\left(\frac{\lambda^2}{2} x_{1i}^2 + \lambda x_{1i}x_{2i} + x_{2i}^2\right) + \int_0^{x_{1i}} f_i(\sigma_i)\,d\sigma_i \tag{6.48}$$

$\dot{v}_i(\underline{x}_i)$ along (6.46) is evaluated as

$$\dot{v}_i(\underline{x}_i) = -\frac{\lambda}{2}\frac{f_i(x_{1i})}{x_{1i}}x_{1i}^2 - \frac{\lambda}{2}x_{2i}^2 \tag{6.49}$$

Since the nonlinearity $f_i(x_{1i})$ is in a region around the origin, we get an upper bound on $\frac{f_i(x_{1i})}{x_{1i}}$ as β where

$$\beta > \left.\frac{\partial f_1(x_{1i})}{\partial x_{1i}}\right|_{x_{1i}=0} = (1/M_{eq})\cos(\delta_{in}^s+\psi_{in})$$

Hence

$$\dot{v}_i(\underline{x}_i) \le -\underline{x}_i^T\underline{Q}\underline{x}_i, \text{ where } \underline{Q} = \text{Diag}(\lambda\beta/2,\lambda/2)$$
$$\le -\sigma_i\|\underline{x}_i\|^2 \tag{6.50}$$

where σ_i is $\min(\lambda\beta/2,\lambda/2)$.

The region of stability of the free subsystem is given using Eq. (4.47) and the equivalences (6.47) as

$$v_i(\underline{x}_i) < [\pi - 2(\delta_{in}^s+\psi_{in})]^2\frac{\lambda^2}{4} + \frac{2}{M_{eq}}\int_0^{\pi-2(\delta_{in}^s+\psi_{in})} f_i(\sigma_i)d\sigma_i \tag{6.51}$$

i.e. $v_i(\underline{x}_i) < v_{oi}$

6.7.2 Stability of Composite System

The weighted sum scalar Lyapunov function is

$$V(\underline{x}) = \sum_{i=1}^{n-1}\alpha_i v_i(\underline{x}_i)$$
$$\dot{V}(\underline{x}) = \sum_{i=1}^{n-1}\{\alpha_i[\dot{v}_i(\underline{x}_i)|_{(6.46)} + [\text{grad } v_i(\underline{x}_i)]^T\underline{h}_i(\underline{x})]\} \tag{6.52}$$

Applying the obvious inequalities

$$|x_{1i} - x_{1k}| \le |x_{1i}| + |x_{1k}|$$
$$\text{and} \quad |x_{1i}| < \|\underline{x}_i\| \quad \text{and} \quad |x_{2i}| \le \|\underline{x}_i\| \tag{6.53}$$

and another inequality suggested in Ref. [7]

$$\cos(x_{1i}-x_{ki}+\delta^o_{ik}-\theta_{ik}) - \cos(\delta^o_{ik}-\theta_{ik})$$

$$\leq |\sin(\delta^o_{ik}-\theta_{ik})||x_{1i} - x_{1k}| \tag{6.54}$$

We majorize the second term in the right-hand side of (6.52).

$$\underline{h}_i(\underline{x}) = \begin{bmatrix} 0 \\ h_{i2}(\underline{x}) \end{bmatrix} \tag{6.55}$$

where
$$h_{i2}(\underline{x}) = M_n^{-1} \sum_{\substack{j=1 \\ \neq i}}^{n-1} A_{nj}[\cos(\delta_{nj}-\phi_{nj}) - \cos(\delta^s_{nj}-\phi_{nj})]$$
$$- M_i^{-1} \sum_{\substack{j=1 \\ \neq i}}^{n-1} A_{ij}[\cos(\delta_{ij}-\phi_{ij}) - \cos(\delta^s_{ij}-\phi_{ij})] \tag{6.56}$$

Now
$$[\text{grad } v_i(\underline{x}_i)]^T \underline{h}_i(\underline{x}) = (\tfrac{\lambda}{2} x_{1i}+x_{2i})h_{i2}(\underline{x})$$

$$\leq (1 + \tfrac{\lambda}{2}) ||\underline{x}_i|| \cdot ||h_{i2}(\underline{x})||$$

$$\leq (1 + \tfrac{\lambda}{2})\{M_i^{-1} \sum_{\substack{j=1 \\ \neq i}}^{n-1} A_{ij}|\sin(\delta^s_{ij}-\phi_{ij})||x_{1i}-x_{1j}| \tag{6.57}$$
$$+ M_n^{-1} \sum_{\substack{j=1 \\ \neq i}}^{n-1} A_{nj}|\sin(\delta^s_{nj}-\phi_{nj})||x_{1j}|\}\cdot||\underline{x}_i||$$

$$\leq (1 + \tfrac{\lambda}{2})\{M_i^{-1} \sum_{j=1}^{n-1} A_{ij}|\sin(\delta^s_{ij}-\phi_{ij})|\cdot||\underline{x}_i||^2$$
$$+ M_i^{-1} \sum_{\substack{j=1 \\ \neq i}}^{n-1} A_{ij}|\sin(\delta^s_{ij}-\phi_{ij})|\cdot||\underline{x}_i||\cdot||\underline{x}_j|| \tag{6.58}$$
$$+ M_n^{-1} \sum_{\substack{j=1 \\ \neq i}}^{n-1} A_{nj}|\sin(\delta^s_{nj}-\phi_{nj})|\cdot||\underline{x}_i||\cdot||\underline{x}_j||\}$$

Hence from (6.52) and using (6.58)

$$\dot{V}(\underline{x}) \leq -\sum_{i=1}^{n-1} \alpha_i \sigma_i \|\underline{x}_i\|^2 + \sum_{i=1}^{n-1} \sum_{j=1}^{n-1} \alpha_i a_{ij} \|\underline{x}_i\| \cdot \|\underline{x}_j\| \quad (6.59)$$

where

$$a_{ii} = \frac{(1+\lambda/2)}{M_i} \sum_{\substack{j=1\\ \neq i}}^{n-1} A_{ij} |\sin(\delta_{ij}^s - \phi_{ij})|$$

$$a_{ij} = (1 + \frac{\lambda}{2}) \{M_n^{-1} A_{nj} |\sin(\delta_{nj}^s - \phi_{nj}) \quad (6.60)$$

$$+ M_n^{-1} A_{ij} |\sin(\delta_{ij}^s - \phi_{ij})|\} \qquad i \neq j$$

By the majorization process, we naturally have $a_{ij} \geq 0$ for $i \neq j$. Also, $a_{ii} \geq 0$. We can now apply any of the conditions (i), (ii) or (iii) of Theorem 6.2 to ensure asymptotic stability of the composite system, the $(n-1)\times(n-1)$ test matrix being $\underline{D} = [d_{ij}]$ where

$$d_{ij} = \begin{cases} \sigma_i - a_{ii} & i=j \\ -a_{ij} & i \neq j \end{cases} \quad (6.61)$$

In particular, we utilize the row dominance property, i.e., there must exist $\alpha_i > 0$ $(i=1,2,\ldots,n-1)$ such that

$$\alpha_i(\sigma_i - a_{ii}) - \sum_{\substack{j=1\\ \neq i}}^{n-1} \alpha_j a_{ij} > 0 \qquad i=1,2,\ldots,n-1 \quad (6.62)$$

This implies $\alpha_i(\sigma_i - a_{ii}) > 0$ and since $\sigma_i > 0$, we have $\sigma_i > a_{ii}$, i.e.

$$\sigma_i > \frac{(1+\lambda/2)}{M_i} \sum_{\substack{j=1\\ \neq i}}^{n-1} A_{ij} |\sin(\delta_{ij}^s - \phi_{ij})| \quad (6.63)$$

Equation (6.63) indicates that σ_i which is a function of damping as well as the pre-fault state of the individual subsystems should be such as to overcome the factors arising through the network interconnections Y_{ij}. This equation also reveals that the weaker the value of Y_{ij} $(i \neq j)$ the easier it is for the $\sigma_i - a_{ii}$ to be > 0 which is only a sufficient condition. It also means that the decomposition should be performed in such a manner that the resultant subsystems are weakly coupled [3]. In the decomposition based on the internal node description, Y_{ij}'s are fixed and hence there is very little flexibility. Decomposition techniques based on structure preserving models [15] might be helpful.

6.7.3 Numerical Example [13]

The four machine example of Fig. 6.1 is taken from El-Abiad and Nagappan [18] and modified to provide uniform damping characteristics. Tables 6.1-6.6 give the pertinent data. Table 6.7 shows that four different fault locations are examined. For example, in Test No. 1 we assume that a 3-phase fault occurs near bus 8 and that line 8-9 is tripped after fault is cleared. We assume λ = .001 and β = 1.0. The numerical results are adapted from Chen and Schinzinger [13].

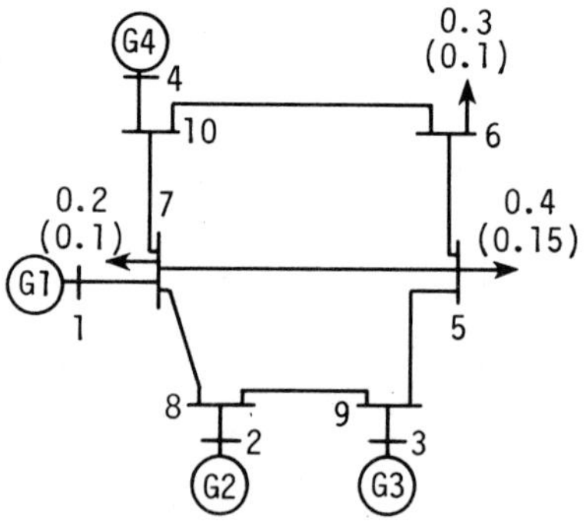

Fig. 6.1 Single line diagram of four machine power system.

From Bus	To Bus	P.U. Impedance
5	6	0.2 + j0.4
5	7	0.2 + j0.5
5	9	0.1 + j0.3
6	10	0.1 + j0.15
7	8	0.1 + j0.5
7	10	0.05 + j0.2
8	9	0.2 + j0.8

Table 6.1 Line Constants

Generator Number	MVA Capacity	X_d' in P.U.	M in P.U.	D in P.U.
1	15	1.0	1,130.0	1.13
2	40	0.5	2,260.0	2.26
3	30	0.4	1,508.0	1.508
4	100	0.004	75,350.0	75.35

NOTES: 1. $M_i = 4\pi f H_i$ P.U. - rad^2/rad
2. $D_i = 2\pi f d_i$ P.U. - rad/rad
3. H_i = inertia constant in P.U./sec
4. d_i = damping coefficient in P.U.-sec

Table 6.2 Machine Constants

	1	2	3	4
1	$0.88\angle -88.1°$	$0.124\angle 82.1°$	$0.065\angle 82.4°$	$0.658\angle 91.1°$
2	$0.124\angle 82.1°$	$0.873\angle -83.2°$	$0.064\angle 88.2°$	$0.655\angle 96.8°$
3	$0.065\angle 82.4°$	$0.064\angle 88.2°$	$1.014\angle -75.5°$	$0.754\angle 99.0°$
4	$0.658\angle 91.1°$	$0.655\angle 96.8°$	$0.754\angle 99.0°$	$2.447\angle -69.7°$

Table 6.3 Reduced Bus Admittance Matrix for Post-fault System (line 8-9 tripped)

	1	2	3	4
1	$0.89\angle{-88.3^\circ}$	0	$0.044\angle{81.0^\circ}$	$0.581\angle{90.6^\circ}$
2	0	$2.0\angle{-90^\circ}$	0	0
3	$0.044\angle{81.0^\circ}$	0	$1.354\angle{-80.8^\circ}$	$0.533\angle{97.9^\circ}$
4	$0.581\angle{90.6^\circ}$	0	$0.533\angle{97.9^\circ}$	$2.9\angle{-69.6^\circ}$

Table 6.4 Reduced Bus Admittance Matrix for Fault-on System (3 Phases Short-Circuited as Bus 8 End of Line 8-9)

Generator Number	$E_i\angle{\delta_{in}}$
1	$1.057\angle{5.30^\circ}$
2	$1.155\angle{11.08^\circ}$
3	$1.095\angle{5.48^\circ}$
4	$1.000\angle{0.08^\circ}$

Table 6.5 Internal Voltage of Pre-Fault System

Generator Number	$E_i\angle{\delta_{in}}$
1	$1.057\angle{5.69^\circ}$
2	$1.152\angle{14.39^\circ}$
3	$1.095\angle{2.27^\circ}$
4	$1.000\angle{0.08^\circ}$

Table 6.6 Internal Voltage for Post-Fault System (line 8-9 tripped)

Test Number		1	2	3	4
Fault Location		Line 8-9 at Bus 8 Side	Line 8-9 at Bus 9 Side	Line 5-7 at Bus 7 Side	Line 5-7 at Bus 5 Side
Line Tripped		Line 8-9	Line 8-9	Line 5-7	Line 5-7
Critical Switching Time in sec	Criterion 1	0.36	0.38	0.25	0.30
	Criterion 2	0.36	0.38	0.30	0.37
	Runge-Kutta Method	0.36	0.40	0.35	0.44

Table 6.7 Critical Switching Time for Four-Machine Systems

With machine 4 as reference and letting $\alpha_i = 1$, i=1,2,3, the following $\underline{D}$ matrix is obtained using (6.60) with $\sigma_i = .0005$.

$$\underline{D} = \begin{bmatrix} 3.013 & -0.733 & -0.586 \\ -1.431 & 4.010 & -0.623 \\ -0.757 & -0.456 & 3.976 \end{bmatrix} \times 10^{-4} \tag{6.64}$$

It may bè verified that $\underline{D}$ is indeed an M-matrix. Hence the equilibrium is asymptotically stable. Before optimizing the α_i's, we compute v_{oi} (i=1,2,3) the stability regions of the three free subsystems. To illustrate, the free subsystem for machine 1 is

$$\frac{dx_{1i}}{dt} = x_{2i} \tag{6.65}$$

$$\frac{dx_{2i}}{dt} = -.001\ x_{21} - 6.251\times10^{-4}[\sin(\sigma_1+0.0786) - 0.0786]$$

$$\sigma_1 = [1 \quad 0]\begin{bmatrix} x_{1i} \\ x_{2i} \end{bmatrix}$$

The nonlinearity $f_1(\sigma_1)$ satisfies the condition $\sigma_1 f_1(\sigma_1) > 0$ in the region $-\pi - 2(\delta^s_{in}+\psi^s_{in}) < \sigma_1 < \pi - 2(\delta^s_{in}+\psi_{in})$, i.e. $-3.298 < \sigma_1 < 2.984$. Using (6.47) the stability region v_{01} is computed as 1.1×10^{-3}. Similarly the region for machines 2 and 3 are computed as $v_{02} = 5.477\times10^{-4}$ and $v_{03} = 9.281\times10^{-4}$ respectively.

6.7.4 Optimizing α_i's and Computing the Region of Attraction

We use the results of section 6.6 to optimize the region of attraction by an appropriate choice of weightages α_i's. Using Criterion 1 gives

$$\text{Min } z = (3.013\alpha_1+4.01\alpha_2+3.976\alpha_3)10^{-4} \tag{6.67}$$

Subject to

$$\begin{aligned} &\alpha_1 + \alpha_2 + \alpha_3 = 1 \\ &3.013\alpha_1 - 0.733\alpha_2 - 0.586\alpha_3 > 0 \\ &-1.431\alpha_1 + 4.01\alpha_2 - 0.623\alpha_3 > 0 \\ &-0.757\alpha_1 - 0.456\alpha_2 + 3.976\alpha_3 > 0 \end{aligned} \tag{6.68}$$

The solution to the above linear programming problem is found to be $\alpha_1^* = 0.6139$, $\alpha_2^* = -0.2415$, $\alpha_3^* = 0.1446$ and $\min_{i=1,2,3} (\alpha_i^* v_{oi})$ turns out to be 1.32×10^{-4}. Hence the stability region is

$$\sum_{i=1}^{3} \alpha_i^* v_i(x_i) < 1.32\times10^{-4}$$

For this particular post-fault condition, the solution obtained by using Criterion 2 yields the same as α_i^*'s in Criterion 1. The critical switching time determined using the procedure discussed in Chapter V is 0.36 sec. The value of t_{cr} by actual simulation is found to be 0.37 sec. Thus there is a good agreement by both the methods.

Table 6.7 summarizes the list of results at the several fault locations in this four machine system. It appears that Criterion 2 gives the results which are less conservative. However, by both criteria the results agree very well by actual simulation method for tests 1 and 2. The discrepancy is larger for tests 3 and 4. This means that the decomposition technique gives favorable results only for certain faults and not for other ones. This is not a desirable feature and more research is needed to remove this drawback.

6.8 Vector Lyapunov Function Approach Using the Comparison Principle

Historically the vector Lyapunov function method using comparison principle was the first technique to be used for stability analysis of large scale systems via decomposition [1,2]. In one sense, it is a very general approach and has naturally generated a lot of interest in control theory and mathematical circles. However, for most of the specific cases considered thus far, the method reduces to the weighted sum scalar Lyapunov function method where usage of the comparison principle is not required. In power systems, comparison principle based vector Lyapunov function (VLF) was used initially together with the requirement that the free subsystem be exponentially stable [6,7,8]. This condition is, however, stringent so far as power systems are considered and hence the results were extremely conservative. Therefore, the method did not result in practical applications to realistic systems, in spite of the relaxation on the conditions on the subsystems [9]. In this section, we briefly review the VLF method and indicate the application to the power system problem. The comparison principle is discussed extensively in the literature [3,4]. Our presentation is along the lines of Michel and Miller [3].

6.8.1 Comparison Principle

Consider a system of k differential equations

$$\dot{\underline{y}} = \underline{H}(\underline{y}), \quad \underline{y}(t_o) = \underline{y}_o \tag{6.69}$$

Usually we assume that (6.69) possesses unique solutions for every $\underline{y}_o \varepsilon R^k$ and for all $t > 0$.

Definition

$\underline{H}(\underline{y})$ is said to be quasi-monotone if for each component H_j $j=1,2,\ldots,k$ the inequality

$$H_j(\underline{y}) \leq H_j(\underline{z})$$

is true whenever $y_i \leq z_i$ for all $i \neq j$ and $y_j = z_j$. If $H(y) = \underline{P}\,\underline{y}$ where $\underline{P}$ is a $k \times k$ matrix, then the quasi-monoticity condition reduces to the condition $p_{ij} \geq 0$ $(i \neq j)$.

Comparison Principle Theorem

If $\underline{r}(t)$ is a continuous function and satisfies

$$\dot{\underline{r}} \leq \underline{H}(\underline{r})$$

such that $\underline{r}(t_o) \leq \underline{y}(t_o)$, then a necessary and sufficient condition for the inequality

$$\underline{r}(t) \leq \underline{y}(t) \quad \text{for} \quad t \geq t_o \tag{6.71}$$

to hold is that $\underline{H}$ be quasi-monotone.

In the case $\underline{H}(\underline{V}) = \underline{P}\ \underline{V}$, the condition reduces to $p_{ij} \geq 0$ $(i \neq j)$. For the linear case it also follows that if $\underline{P}$ has eigenvalues whose real part < 0, then $\underline{y}(t)$ and $\underline{r}(t) \to 0$ as $t \to \infty$. This, together with the property that $p_{ij} \geq 0$ $(i \neq j)$ is equivalent to $-\underline{P}$ being an M-matrix.

We now state the theorem for a.s.i.l. of the system (6.23) using the comparison principle and VLF.

Theorem 6.3

Let the free subsystems (6.24) have the Lyapunov function $v_i(\underline{x}_i)$ and let

$$\dot{v}_i(\underline{x}_i)\Big|_{(6.24} \leq -\beta_i\ v_i(\underline{x}_i), \quad \beta_i > 0$$

Define the vector Lyapunov function $\underline{V}(\underline{x})$ as $(v_1(\underline{x}_1),\ldots,v_k(\underline{x}_k))^T$. Now

$$\begin{aligned}\dot{v}_i(\underline{x}_i)\Big|_{(6.23)} &= [\underline{\nabla v}_i(\underline{x}_i)]^T[\underline{f}_i(\underline{x}_i)+\underline{g}_i(\underline{x}_i)] \\ &= -\beta_i\ v_i(\underline{x}_i) + \underline{\nabla v}_i(\underline{x}_i)^T\underline{g}_i(\underline{x}_i)\end{aligned} \tag{6.72}$$

Let us assume that there exists a k×k matrix $\underline{P} = [p_{ij}]$, $p_{ij} \geq 0$ $i \neq j$ such that

$$\dot{v}_i(\underline{x}_i)\Big|_{(6.23)} \leq \sum_{j=1}^{k} p_{ij}\ v_i(\underline{x}_i) \tag{6.73}$$

Note that β_i is absorbed in p_{ii}. Hence

$$\dot{\underline{V}}(\underline{x}) \leq \underline{P}\ \underline{V}(\underline{x})$$

If the matrix $\underline{P}$ has eigenvalues with negative real parts, the equilibrium of the composite system is asymptotically stable.

Proof:' The proof involves using Theorem 6.1 concerning weighted sum scalar Lyapunov function. The two assumptions, namely that $\underline{P}$ has eigenvalues with negative real part and $p_{ij} \geq 0$ $(i \neq j)$, imply that $-\underline{P}$ is an M-matrix. From the properties of M-matrices in Sec. (6.5), we have the equivalent computable condition

$$(-1)^{\ell} \begin{vmatrix} p_{11} & \cdots & p_{1\ell} \\ \\ p_{\ell 1} & \cdots & p_{\ell\ell} \end{vmatrix} > 0 \qquad \ell=1,2,\ldots,k \tag{6.74}$$

Alternatively, we can use property (ii) and equate η_j's with α_j's. Thus, we can postulate the existence of a vector $\underline{\alpha} = (\alpha_1,\ldots,\alpha_k)^T > 0$ such that $\underline{\alpha}^T\underline{P} < 0$. We can define the scalar Lyapunov function

$$V(\underline{x}) = \sum_{\alpha=1}^{k} \alpha_i v_i(\underline{x}_i) \tag{6.75}$$

$V(\underline{x})$ is positive definite and

$$\dot{V}(\underline{x}) = \sum_{\alpha=1}^{k} \alpha_i \dot{v}_i(\underline{x}_i) = \underline{\alpha}^T \, \dot{\underline{V}}(\underline{x})$$

$$\leq \underline{\alpha}^T\underline{P}\ \underline{V}(\underline{x}) \tag{6.76}$$

Since $\underline{\alpha}^T\underline{P} < 0$, $\underline{\alpha}^T\underline{P}\ \underline{V}$ is < 0 and hence $\dot{V}(\underline{x}) < 0$. Hence the origin is asymptotically stable. We have also thus shown that in the case of a linearized version of the comparison principle, the VLF and weighted sum scalar Lyapunov function are really equivalent.

6.9 Power System Application of Vector Lyapunov Function Method

It has been pointed out that the VLF method was the first one to be applied to power systems [6] and hence there is a good amount of literature using this method. To construct the $\underline{P}$ matrix which satisfies the inequality $\dot{\underline{V}}(\underline{x}) \leq \underline{P}\ \underline{V}(\underline{x})$, it is generally necessary to have $v_i(\underline{x}_i)$ which are quadratic functions of $\underline{x}_i$. Hence, it poses a problem in power system application where $v_i(\underline{x}_i)$ consists of a quadratic plus an integral of the nonlinearity. Ways of circumventing this difficulty often resulted in very conservative results. We briefly sketch a number of approaches to the problem.

6.9.1 Approach Based on Exponential Stability of Subsystems [20,7]

The uniform damping case is discussed. With each subsystem we associate a Lyapunov function $\hat{v}_i(\underline{x}_i) \triangleq v_i^{1/2}(\underline{x}_i)$ where $v_i(\underline{x}_i)$ is given by (6.48). Each subsystem is assumed to be exponentially stable with the following estimates on $\hat{v}_i(\underline{x}_i)$ and $\dot{\hat{v}}_i(\underline{x}_i)$.

$$\eta_{i1}\|\underline{x}_i\| \le \hat{v}_i(\underline{x}_i) \le \eta_{i2}\|\underline{x}_i\|, \quad \dot{\hat{v}}_i(\underline{x}_i) \le -\eta_{i3}\|\underline{x}_i\| \tag{6.77}$$

$$\|[\underline{\nabla v}_i(\underline{x}_i)]^T \underline{h}_i(\underline{x})\| \le \sum_{j=1}^{k} \xi_{ij}\|\underline{x}_j\| \tag{6.78}$$

The vector Lyapunov function is now $\underline{V}(\underline{x}) = (\hat{v}_1(\underline{x}_1), \hat{v}_2(\underline{x}_2), \ldots, \hat{v}_k(\underline{x}_k))^T$. It can be shown that in the inequality $\dot{\underline{V}}(\underline{x}) \le \underline{P}\ \underline{V}(\underline{x})$, $\underline{P}$ is given by

$$p_{ij} \doteq \begin{cases} -\eta_{i2}^{-1}\eta_{i3} + \xi_{ii}\eta_{i1}^{-1} & i=j \\ \xi_{ij}\eta_{j1}^{-1} & i\neq j \end{cases} \tag{6.79}$$

In (6.77) and (6.78) η_{ik}'s are positive numbers and ξ_{ij}'s are non-negative numbers. Hence $-\underline{P}$ in (6.79) is an M-matrix and if the conditions (6.74) are satisfied, the equilibrium of (6.23) is asymptotically stable.

6.9.2 Region of Attraction [7]

Let the region of attraction of the free subsystem be defined by

$$v_i(\underline{x}_i) < v_{oi} \tag{6.80}$$

Define $\hat{v}_i^o = v_{oi}^{1/2}$. Following Weissenberger [5], the region of attraction is defined by

$$\nu(\underline{x}) < \gamma \tag{6.81}$$

where $\nu(\underline{x}) = \max_{1\le i\le k} |\hat{v}_i(\underline{x}_i)|/p_i$

and $\gamma = \min_{1\le i\le k} (\hat{v}_i^o/p_i)$

$\underline{p}$ is computed using the properties of $-\underline{P}$ which is an M-matrix (see Sec. 6.5) as follows: Since $-\underline{P}$ is an M-matrix, $-\underline{P}^{-1}$ is non-singular and all elements of $-\underline{P}^{-1}$ are non-negative (all diagonal elements of $-\underline{P}^{-1}$ are positive). Hence, given any vector $\underline{r} > 0$, there exists a $\underline{p} > 0$ defined by

$$\underline{p} = -\underline{P}^{-1}\underline{r}$$

The application of this method for a power system is briefly sketched below for the uniform damping case given by (6.45). The subsystem Lyapunov function $v(\underline{x}_i)$ satisfies the following inequalities which can be deduced from (6.48)

$$\underline{x}_i^T \underline{H}_i \underline{x}_i \leq v_i(\underline{x}_i) \leq \underline{x}_i^T \underline{H}_i \tilde{\underline{x}}_i \tag{6.83}$$

where

$$\underline{H}_i = \frac{1}{2}\begin{bmatrix} \frac{\lambda^2}{2} & \frac{\lambda}{2} \\ \frac{\lambda}{2} & 1 \end{bmatrix} \quad \text{and} \quad \tilde{\underline{H}}_i = \underline{H}_i + \frac{1}{2}\beta\, \underline{c}_i \underline{c}_i^T$$

where β is the slope of $f_i(\sigma_i)$ at $\sigma_i = 0$. The bound on $\dot{v}_i(\underline{x}_i)$ is given by (6.50) as

$$\dot{v}_i(\underline{x}_i) \leq -\underline{x}_i^T \underline{Q}_i \underline{x} \tag{6.84}$$

Since $\hat{v}_i(\underline{x}_i) = v_i^{1/2}(\underline{x}_i)$, the numbers $\eta_{i1}, \eta_{i2}, \eta_{i3}$ are chosen as

$$\eta_{i1} = \lambda_m^{1/2}(\underline{H}_i), \quad \eta_{i2} = \lambda_M^{1/2}(\tilde{\underline{H}}_i),$$

$$\eta_{i3} = \frac{\lambda_m(\underline{Q}_i)}{2\lambda_M^{1/2}(\tilde{\underline{H}}_i)} \tag{6.85}$$

where λ_m and λ_M are the minimum and maximum eigenvalues of the indicated matrices. The ξ_{ij}'s are computed using a procedure similar to the computation of α_{ij}'s in (6.60). Details are to be found in Ref. [7].

This methodology has been extended to the non-uniform damping case in Ref. [8] with the stability region defined as

$$\sum_{i=1}^{n-1} \alpha_i v_i(\underline{x}_i) < \gamma$$

where α_i's are chosen to satisfy (6.39) with $\underline{D} = \underline{P}$ i.e. $\underline{P}^T\underline{A} + \underline{A}\ \underline{P}$ is positive definite with $\underline{A} = \mathrm{Diag}(\alpha_i)$. γ is computed as $\min_{1 \leq i \leq n-1} (\alpha_i v_i^o)$. Therefore, the method essentially reduces to that of the weighted sum scalar Lyapunov function method. However, the majorizing process is different in the two cases. The VLF majorization seems to be more restrictive than the WSSLF method. This perhaps explains why the results are better in the latter case.

6.9.3 Combined VLF and Weighted Sum Scalar Lyapunov Function Approach [9]

In the weighted sum scalar Lyapunov function method, we chose majorization on $v_i(\underline{x}_i)$ to be $\dot{v}_i(\underline{x}_i) < -\sigma_i \|\underline{x}_i\|^2$. Instead of $\|\underline{x}_i\|^2$, we can choose other functions such as positive definite functions $u_i(\underline{x}_i)$ so that $\dot{v}_i(\underline{x}_i) \leq -\sigma_i\ u_i^2(\underline{x}_i)$. The analysis can proceed along the same lines as in Sec. 6.3. In the weighted sum scalar Lyapunov function method (Theorem 6.1), we imposed no sign restrictions on a_{ij}. In the case of power systems, the very nature of majorization yields $a_{ij} \geq 0$ $i \neq j$. Instead of considering σ_i and a_{ij} separately, we can assume what is called a "one-shot" aggregation by assuming the existence p_{ij} such that

$$\dot{v}_i(\underline{x}_i) = [\mathrm{grad}\ v_i(\underline{x}_i)]^T[\underline{f}_i(\underline{x}_i) + \underline{h}_i(\underline{x}_i)]$$

$$\leq \sum_{j=1}^{k} p_{ij}\ u_i(\underline{x}_i)u_j(\underline{x}_j) \tag{6.86}$$

We can now restate the stability theorem, 6.1, as follows [9].

<u>Theorem 6.4</u>

Given $v_i(\underline{x}_i)$ and $u_i(\underline{x}_i) > 0$ and $\dot{v}_i(\underline{x}_i)$ satisfying (6.86), if there exists a diagonal matrix $\underline{B} = (\alpha_1,\ldots,\alpha_k)^T > 0$ such that

$\underline{P}^T\underline{B} + \underline{B}\ \underline{P}$ is negative definite, then the origin of (6.23) is asymptotically stable.

In the above theorem, p_{ij} may be of any sign. A closer look at this theorem reveals that it is equivalent to Theorem 6.1 with

$$u_i(\underline{x}_i) = \|\underline{x}_i\|$$

$$p_{ij} = \begin{cases} -\sigma_i + a_{ii} & i=j \\ a_{ij} & i \neq j \end{cases}$$

Furthermore, $\underline{P}^T\underline{B} + \underline{B}\ \underline{P} = 2\underline{S}$. Hence, the two theorems are entirely equivalent. If in addition $\underline{P}$ is an M-matrix, then properties of the M-matrix can be used to get optimum value of $\underline{\alpha}$.

The application of the preceding theorem to power systems has been done in Ref. [10] whose main features are

(i) In the power system model, asynchronous electromechanical damping in addition to mechanical damping has been taken into account.

(ii) Governor action is incorporated for each machine.

(iii) The decomposition is done along the lines discussed in 6.2.1(ii), i.e. there is no overlapping decomposition.

(iv) The region of attraction is obtained as the union of the two regions, one obtained by weighted sum scalar Lyapunov function method and the other as in VLF method.

However, the computation results reported are quite conservative. The main reason is the type of majorization used which removes information about the fine physical structure of the power system and the absence of any optimal choice of α_i's.

Conclusions

In this chapter, we have presented a systematic methodology of stability analysis by large scale power systems by decomposition. As shown, the weighted sum scalar Lyapunov function approach is equivalent to the vector Lyapunov function approach in most applications considered so far. While applying the VLF approach, the requirement of exponential stability gives rise to overly conservative results.

Further research in large scale power system stability should be directed towards improved methods of decomposition retaining the physical integrity of the system and optimizing the region of attraction.

References

1. Bellman, R., "Vector Lyapunov Functions", SIAM Journal of Control, Vol. 1, 1962, pp. 32-34.

2. Bailey, F. N., "The Application of Lyapunov's Second Method to Interconnected Systems", SIAM Journal of Control, Vol. 3, 1966, pp. 443-462.

3. Michel, A. N. and Miller, R. K., "Qualitative Analysis of Large Scale Dynamical Systems", (Book) Academic Press, New York, 1977.

4. Siljak, D. D., "Large Scale Dynamic Systems - Stability and Structure", North Holland, New York, 1978.

5. Weissenberger, S., "Stability Regions of Large Scale Systems", Automatica, Vol. 9, 1979, pp. 653-663.

6. Pai, M. A. and Narayana, C. L., "Stability of Large Scale Power Systems", Proceedings of 6th IFAC World Contress, Boston, 1975, Paper 31.6, pp. 1-10.

7. Jocic, Lj. B., Ribbens-Pavella, M. and Siljak, D. D., "Multimachine Power Systems: Stability, Decomposition and Aggregation", IEEE Trans., Vol. AC-23, 1978, pp. 325-332.

8. Jocic, Lj. B. and Siljak, D. D., "Decomposition and Stability of Multimachine Power Systems", Proceedings of 7th IFAC Congress, Helsinki, 1978, pp. 21-26.

9. Gruijc, Lj. T. and Ribbens-Pavella, M., "Relaxed Large Scale Systems Stability Analysis Applied to Power Systems", Proc. 7th IFAC World Congress, Helsinki, 1978.

10. Ribbens-Pavella, M., Gruijc, Lj. T. and Sabatel, J., "Direct Methods for Stability Analysis of Large Scale Power Systems", Proc. of IFAC Symposium on 'Computer Applications on Large Scale Power Systems', New Delhi, India, Aug. 16-18, 1979.

11. Gruijc, Lj. T., Darwish, M. and Fantin, J., "Coherence, Vector Lyapunov Functions and Large Scale Power Systems", International Journal of Control, Vol. 10, 1979, pp. 351-362.

12. Araki, M., Mohsen-Metwally, M. and Siljak, D. D., "Generalized Decomposions for Transient Stability Analysis of Multi-machine Power Systems", Proceedings of JACC, San Francisco, Paper No. TA3-B, 1980.

13. Chen, Y. K. and Schinzinger, R., "Lyapunov Stability of Multimachine Power Systems Using Decomposition-Aggregation Method", Paper A-80-036-4, IEEE Winter PES Meeting, New York, Feb. 1980.

14. Vittal, V., "Lyapunov Stability Analysis of Large Scale Power Systems", M. Tech. Thesis, I.I.T., Kanpur, July 1979.

15. Bergen, A. R. and Hill, D. J., "A Structure Preserving Model for Power System Stability Analysis", IEEE Trans., Vol, PAS-100, Jan. 1981, pp. 25-35.

16. Pai, M. A., "Stability of Large Scale Interconnected Power Systems", Proc. IEEE International Conference on Circuits and Computers, Oct. 1-3, 1980, Port Chester, New York, pp. 413-416.

17. Michel, A. N., "Stability Analysis of Interconnected Systems", SIAM J. Control, Vol. 12, No. 3, Aug. 1974, pp. 554-579.

18. Michel, A. N. and Porter, D. W., "Stability of Composite Systems", Fourth Asilomar Conference on Circuits and Systems, Monterey, California, 1970.

19. El-Abiad, A. H. and Nagappan, K., "Transient Stability Region for Multi-machine Power Systems", IEEE Trans., Vol. PAS-85, No. 2, Feb. 1966, pp. 169-178.

20. Siljak, D. D., "Stability of Large Scale Systems", Proceedings of the 5th IFAC World Congress, Part IV, June 1972, Paris, France.

Chapter VII

CONCLUSIONS

7.1 Introduction

Although it appeared by the mid-70's that Lyapunov's method for power system stability analysis may remain largely as a theoretical development, the picture has brightened considerably as a result of recent research efforts in the computation of stability regions. This was discussed at length in Chapter V where fault location was taken into account in computing t_{cr}. Thus the direct method of stability analysis of power system will continue to be an active area of research in the future with greater emphasis being put on application and on-line implementation for transient security evaluation. Three broad thrust areas of future research can be identified and we will briefly review the work in each one of them.

7.2 Structure Preserving Models

The stability analysis so far has been done on the baiss of converting the loads into constant impedances and obtaining an internal bus description after eliminating all the physical buses. Besides modelling the loads inaccurately, this approach masks the topology of the network, preventing a better understanding of energy and power transfer during the disturbance period. An effort to move away from this viewpoint was first proposed by Bergen and Hill [1]. Their model includes frequency dependence of the loads and retains the topological structure of the transmission network. There is a first order differential equation for the phase angle for each physical bus and functionally the equations are of the form

$$\dot{\underline{x}} = \underline{F}(\underline{x},\underline{y}) \tag{7.1}$$

$$\varepsilon\dot{\underline{y}} = G(\underline{x},\underline{y}) \tag{7.2}$$

where $\underline{x}$ is the state vector associated with the rotor angles and speed deviations and $\underline{y}$ is the state vector associated with phase angles of the physical buses. $\varepsilon > 0$ is a small parameter identified with the damping coefficient at the loads. Equations (7.1) and (7.2) can be cast in the Luré form and the multi-variable Popov Criterion can then be applied [1,2]. While the theoretical development is well advanced, its application to practical systems is still lagging. The main difficulty seems to be in terms of choosing physically realistic parameters for the load model. Using the philosophy of structure preserving models, voltage dependent non-linear static loads have been incorporated and demonstrated successfully on a practical system by Athay and Sun [3].

Equations (7.1) and (7.2) are in the singularly perturbed form and with $\varepsilon = 0$, we obtain a degenerate system or an implicit set of differential equations whose stability is difficult to investigate. Sastry and Varaiya [4] have examined the local stability of the equilibrium points of both the degenerate and the singularly perturbed system. They have also used the formulation for the hierarchial stability analysis and alert state steering control of interconnected power systems. Practical interpretation and validation of some of these ideas yet remain to be done.

It is therefore clear that structure preserving models open up altogether a new area of research for power engineers. Since transmission lines are modelled as such, the line resistance can be neglected in comparison to the reactance thus avoiding the path dependent integral terms in the Lyapunov function. Another advantage of retaining the network structure is the ability to incorporate HVDC links in the stability analysis. In Ref. [5], the concept of distribution factors is used to reflect both the loads and power in the DC link at the internal nodes of the generators as additional power injections. The transmission network is then reduced to the internal buses after neglecting the transmission line resistances. A modified energy function approach is then applied to the system containing the machines and the DC link dynamics. Results of t_{cr} are quite encouraging using this method and extension to systems containing multi-terminal HVDC links is a possibility.

7.3 Stability by Decomposition

In Chapter VI, the analysis was based on the pairwise decompsition of a large scale power system. This does not take advantage of apriori information about the system such as weak coupling, strongly connected machines, etc. Araki, Metwally and Siljak [6] have proposed a generalized decomposition where more than two machines may be included in a subsystem. However, it is still necessary to include the reference machine in each subsystem. The analysis is done on the basis of the internal node description as in Chapter VI. Alternatively if we use a structure preserving model, the possibility of physical decomposition exists (Ref. [16], Chapter IV) using Kron's tearing techniques (Diakoptics) [7]. Here again, practical validation is yet to be demonstrated.

7.4 Security Assessment

In spite of narrowing the gap between the critical clearing time obtained by simulation and Lyapunov's method (using classical model), it is unlikely that Lyapunov's direct method will replace existing simulation methods for planning purposes. The reason is simply that for most of the problems in planning studies, one requires detailed models of the generating unit (i.e. synchronous machine, excitation system and the governor). Lyapunov functions for such complex models are not available at the moment. Hence at best Lyapunov's method will compliment the simulation method in planning studies by filtering out from the large number of cases being studied, those which require detailed scrutiny.

However, it is in the area of transient security monitoring where computational times are very demanding that Lyapunov's method can play a major role. References [37-40] of Chapter I suggest the possible use of Lyapunov function in a security monitoring scheme. However, this subject still requires further research before practically feasible algorithms can be developed.

References

1. Bergen, A. R. and Hill, D. J., "A Structure Preserving Model for Power System Stability Analysis", IEEE Trans. PAS-100, Jan. 1981, pp. 25-35.

2. Pai, M. A., "Some Mathematical Aspects of Power System Stability by Lyapunov's Method" in: Erisman, A. M., Neves, K. W., Dwarakanath, M. H., (Editors) "Electric Power Problems, The Mathematical Challenge", (Book) Society for Industrial and Applied Mathematics (SIAM) Philadelphia, 1980.

3. Athay, T. M. and Sun, D. I., "An Improved Energy Function for Transient Stability Analysis", IEEE International Symposium on Circuits and Systems, Chicago, April 1981.

4. Sastry, S. and Varaiya, P., "Hierarchial Stability and Alert State Steering Control of Power Systems", IEEE Trans. Circuits and Systems, Vol. CAS-27, 1980.

5. Pai, M., Padiyar, K. R. and Radhakrishna, C., "Transient Stability Analysis of Multi-machine AC/DC Power Systems via Energy Function Method", Paper 81 SM 408-4, IEEE PES Summer Power Meeting, Portland, Oregon, July 1981.

6. Araki, M., Metwally, M. M. and Siljak, D. D., "Generalized Decompositions for Transient Stability Analysis of Multi-machine Power Systems", Proceedings of Joint Automatic Control Conference, San Francisco, Aug. 1980.

7. Happ, H. H., "Piecewise Methods and Application to Power Systems", (Book) John Wiley, New York, 1980.

8. Grujic, Lj. T., Darwish, M. and Fantin, J., "Coherence, Vector Lyapunov Functions and Large Scale Power Systems", International Journal of Control, Vol. 10, 1979, pp. 351-362.

9. Mahalanabis, A. K. and Singh, R., "On the Analysis and Improvement of Transient Stability of Multi-machine Power Systems", IEEE Trans. PAS-100, April 1981, pp. 1574-1580.

APPENDIX

SYNCHRONOUS MACHINE MODEL

A.1 Introduction

A knowledge of the synchronous machine model is essential for understanding the various assumptions implicit in the simplfied classical model generally used in transient stability studies. Future research in Lyapunov stability analysis of power systems will undoubtedly try to relax some of these assumptions and accomodate a better model of the synchronous machine. The brief review which follows derives three types of models.

(i) The two axis model with a damper winding in each axis.

(ii) The one axis E_q' model with no damper windings but including the flux decay effects.

(iii) The classical model.

A.2 Model in Park's Variables

The synchronous machine consists of three phase windings on the stator and three windings on the rotor, namely a main field winding along the direct axis and two fictious short circuited damper windings, one each along the direct (d) and quadrature (q) axis respectively (Fig. A.1). We adopt the convention of the d axis leading the q axis following Ref. [1]. A coupled circuit viewpoint of the six windings indicates that the differential equations involve time varying coefficients. It is universal to adopt the well known Park's transformation which converts the three stator windings into two equivalent fictious windings called the d-axis and q-axis windings, moving synchronously with the rotor. After the transformation, the

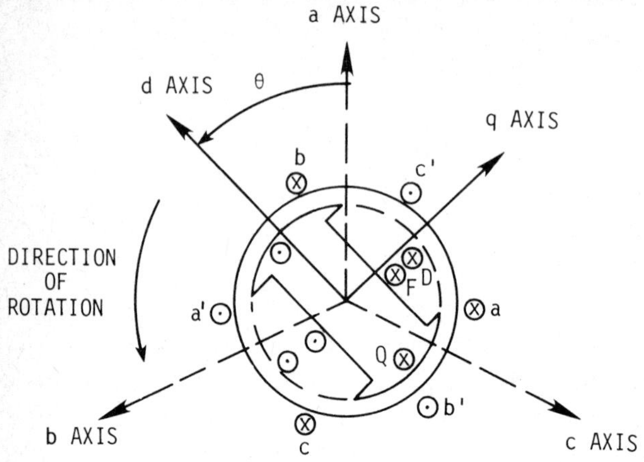

A.1. Pictorial representation of a synchronous machine.

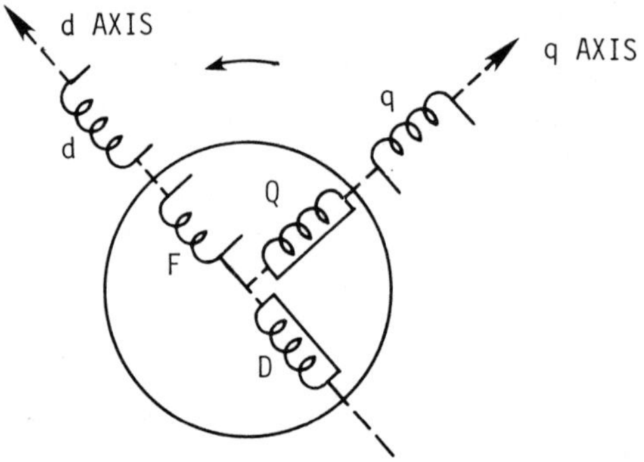

A.2. Equivalent five winding model.

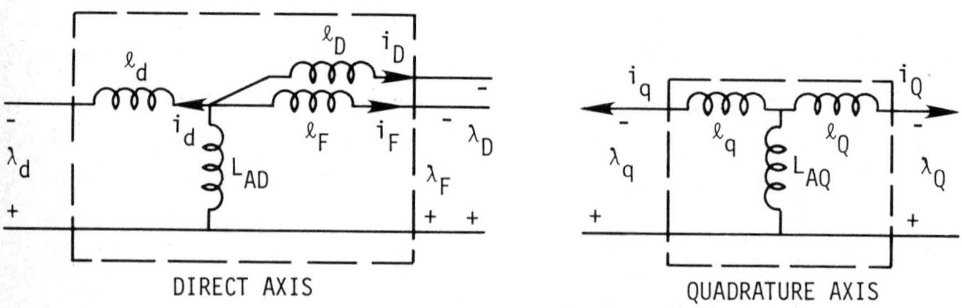

A.3. Multi-port representation of λ-i characteristics in direct and quadrature axis.

differential equations for the five winding model (2 stator and 3 rotor) will now have only time-invariant coefficients. Park's transformation is a power invariant transformation and Ref. [1] contains details of the derivation. We denote the transformed stator windings as d and q axis windings and the rotor field winding as F, the direct axis damper winding as D and the quadrature axis damper winding as Q. Pictorially the five winding model is shown in Fig. A.2. Assuming generator conventions, the Park's equations are given below. These are essentially the Kirchoff's voltage law equations for the five windings and since the stator windings also rotate synchronously with the rotor, there is no mutual coupling between the direct axis and quadratic axis windings. There is, however, the induced voltage in the stator due to rotor action. The equations are:

Stator equations

$$\begin{aligned} v_d &= -ri_d - \frac{d\lambda_d}{dt} - \omega\lambda_q \\ v_q &= -ri_d - \frac{d\lambda_q}{dt} + \omega\lambda_d \end{aligned} \tag{A.1}$$

Rotor Equations

$$\begin{aligned} v_F &= r_F i_F + \frac{d\lambda_F}{dt} \\ 0 &= r_D i_D + \frac{d\lambda_D}{dt} \\ 0 &= r_Q i_Q + \frac{d\lambda_Q}{dt} \end{aligned} \tag{A.2}$$

r's represent the resistance and λ's the flux linkages of the respective windings. The flux current relations in the direct and quadrature axis can be written as

$$\begin{bmatrix} \lambda_d \\ \lambda_F \\ \lambda_D \end{bmatrix} = \begin{bmatrix} L_d & L_{AD} & L_{AD} \\ L_{AD} & L_F & L_{AD} \\ L_{AD} & L_{AD} & L_D \end{bmatrix} \begin{bmatrix} i_d \\ i_F \\ i_D \end{bmatrix} \tag{A.3}$$

$$\begin{bmatrix} \lambda_q \\ \lambda_Q \end{bmatrix} = \begin{bmatrix} L_q & L_{AQ} \\ L_{AQ} & L_Q \end{bmatrix} \begin{bmatrix} i_q \\ i_Q \end{bmatrix} \tag{A.4}$$

where $L_d = L_{AD} + \ell_d$, $L_F = L_{AD} + \ell_F$, $L_D = L_{AD} + \ell_D$, $L_q = L_{AQ} + \ell_q$ and $L_Q = L_{AQ} + \ell_Q$. This assumes a common flux path for each of the two axes. Equations (A.3) and (A.4) can be viewed as a multiport representation as in Fig. A.3. Together Eqs. (A.1) - (A.4) can be given the equivalent circuit representation as in Fig. A.4. The sum of the outputs of the two dependent voltage sources represents the electrical output P_g of the machine, i.e., $P_g = (\lambda_d i_q - \lambda_q i_d)\omega$. From (A.3) and (A.4), we can write the hybrid characterization as [2]

$$\begin{bmatrix} \lambda_d \\ i_F \\ i_D \end{bmatrix} = \begin{bmatrix} L_{MDO} + \ell_d & L_{MDO}\ell_F^{-1} & L_{MDO}\ell_D^{-1} \\ -L_{MDO}\ell_F^{-1} & \ell_F^{-1} - L_{MDO}\ell_F^{-2} & -L_{MDO}\ell_F^{-1}\ell_D^{-1} \\ -L_{MDO}\ell_D^{-1} & -L_{MDO}\ell_F^{-1}\ell_D^{-1} & \ell_D^{-1} - L_{MDO}\ell_D^{-2} \end{bmatrix} \begin{bmatrix} i_d \\ \lambda_F \\ \lambda_D \end{bmatrix} \tag{A.5}$$

$$\begin{bmatrix} \lambda_q \\ i_Q \end{bmatrix} = \begin{bmatrix} L_{MQO} + \ell_q & L_{MQO}\ell_Q^{-1} \\ -L_{MQO}\ell_Q^{-1} & \ell_Q^{-1} - L_{MQO}\ell_Q^{-2} \end{bmatrix} \begin{bmatrix} i_q \\ \lambda_Q \end{bmatrix} \tag{A.6}$$

where $L_{MDO} = \left[\dfrac{1}{L_{AD}} + \dfrac{1}{\ell_F} + \dfrac{1}{\ell_D}\right]^{-1}$ and $L_{MQO} = \left[\dfrac{1}{L_{AQ}} + \dfrac{1}{\ell_Q}\right]^{-1}$

It is customary to define the so-called subtransient inductances in the direct and quadrature axis as $L_d'' = L_{MDO} + \ell_d$ and $L_q'' = L_{MQO} + \ell_q$ respectively. In the absence of damper windings, we have in the direct axis, the transient inductance $L_d' = (1/L_{AD} + 1/\ell_F)^{-1} + \ell_d$ and the steady state inductance in the quadrature axis as $L_q = L_{AQ} + \ell_q$. All these inductances are interpreted in Fig. A.5 as an aid for quick reference. With these preliminaries, we now proceed to develop the model using the hybrid characterization (A.5) and (A.6).

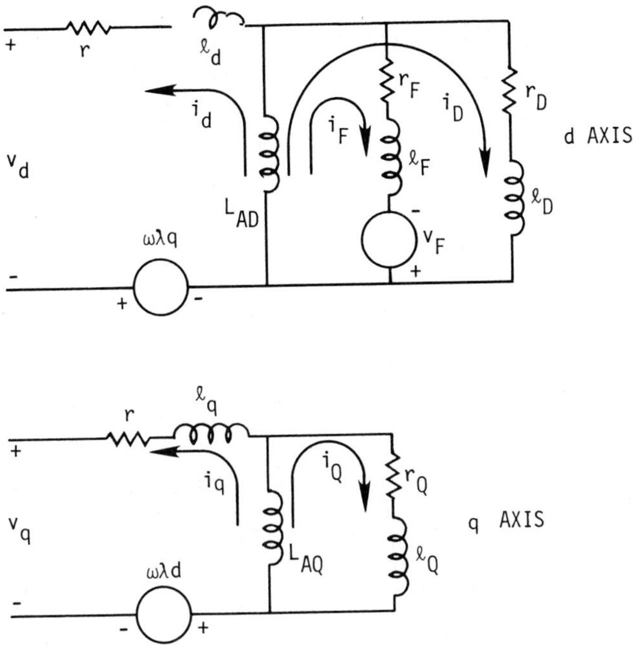

A.4. d and q axis equivalent circuits.

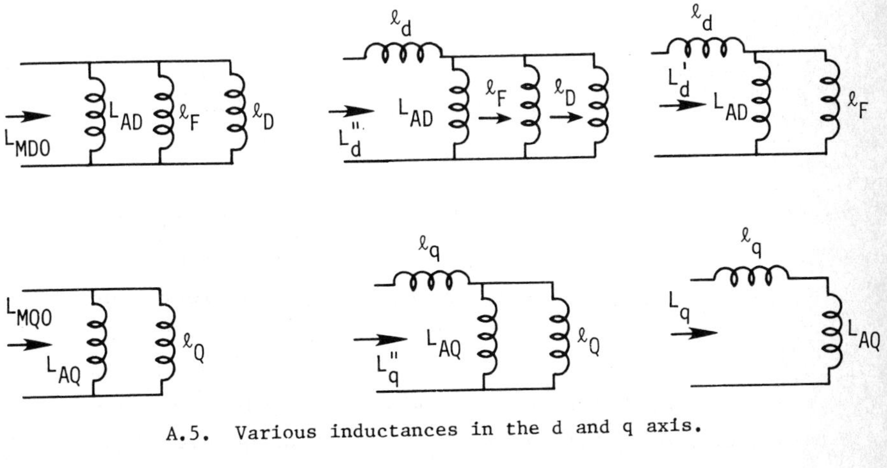

A.5. Various inductances in the d and q axis.

A.3 Hybrid State Space Model (Two axis model)

In transient stability studies, it is usual to neglect the transformer voltages $\frac{d\lambda_d}{dt}$ and $\frac{d\lambda_q}{dt}$ in (A.1) because these are small in magnitude. This makes the equations (A.1) algebraic, i.e.;

Stator equations

$$\begin{aligned} v_d &= -ri_d - \omega\lambda_q \\ v_q &= -ri_q + \omega\lambda_d \end{aligned} \tag{A.7}$$

The rotor equations (A.2) are rearranged as

$$\begin{aligned} \frac{d\lambda_F}{dt} &= -r_F i_F + v_F \\ \frac{d\lambda_D}{dt} &= -r_D i_D \\ \frac{d\lambda_Q}{dt} &= -r_Q i_Q \end{aligned} \tag{A.8}$$

We now substitute the hybrid relations (A.5) and (A.6) in (A.8) for the variables λ_d, λ_q, i_F, i_D and i_Q to get

Stator equations

$$\begin{aligned} v_d &= -ri_d - \omega(L_{MQO}+\ell_q)i_q - \omega L_{MQO}\ell_Q^{-1}\lambda_Q \\ v_q &= -ri_q + \omega(L_{MDO}+\ell_d)i_d + \omega L_{MDO}\ell_F^{-1}\lambda_F + L_{MDO}\ell_D^{-1}\lambda_D \end{aligned} \tag{A.9}$$

Rotor Equations

$$\begin{aligned} \frac{d\lambda_F}{dt} &= r_F L_{MDO}\ell_F^{-1}i_d - r_F(\ell_F^{-1}-L_{MDO}\ell_F^{-2})\lambda_F \\ &\quad + r_F L_{MDO}\ell_F^{-1}\ell_D^{-1}\lambda_D + v_F \\ \frac{d\lambda_D}{dt} &= r_D L_{MDO}\ell_D^{-1}i_d + r_F L_{MDO}\ell_F^{-1}\ell_D^{-1}\lambda_F \\ &\quad - r_D(\ell_D^{-1}-L_{MDO}\ell_D^{-2})\lambda_D \\ \frac{d\lambda_Q}{dt} &= r_Q L_{MQO}\ell_Q^{-1}i_q - r_Q(\ell_Q^{-1}-L_{MQO}\ell_Q^{-2})\lambda_Q \end{aligned} \tag{A.10}$$

Define

$$\frac{1}{\tau''_{fo}} = r_F(\ell_F^{-1} - L_{MDO}\ell_F^{-2}); \quad \frac{1}{\tau''_{do}} = r_D(\ell_D^{-1} - L_{MDO}\ell_D^{-2}) \tag{A.11}$$

and $\frac{1}{\tau''_{qo}} = r_Q(\ell_Q^{-1} - L_{MQO}\ell_Q^{-2})$.

These three time constants are respectively called the field, direct and quadrature axis open circuit time constants. $L''_d = L_{MDO} + \ell_d$ and $L''_q = L_{MQO} + \ell_q$ have been previously defined as the direct axis and quadrature axis subtransient inductances respectively.

Then (A.9) and (A.10) become

Stator equations

$$\begin{aligned} v_d &= -ri_d - \omega L''_q i_q - \omega L_{MQO}\ell_Q^{-1}\lambda_Q \\ v_q &= -ri_q + \omega L''_d i_d + \omega L_{MDO}\ell_F^{-1}\lambda_F + \omega L_{MDO}\ell_D^{-1}\lambda_D \end{aligned} \tag{A.12}$$

Rotor equations

$$\begin{aligned} \frac{d\lambda_F}{dt} &= r_F L_{MDO}\ell_F^{-1} i_d - \lambda_F/\tau''_{fo} + r_F L_{MDO}\ell_F^{-1}\ell_D^{-1}\lambda_D + v_F \\ \frac{d\lambda_D}{dt} &= r_D L_{MDO}\ell_D^{-1} i_d + r_D L_{MDO}\ell_F^{-1}\ell_D^{-1}\lambda_F - \lambda_D/\tau''_{do} \\ \frac{d\lambda_Q}{dt} &= r_Q L_{MQO}\ell_Q^{-1} i_q - \lambda_q/\tau''_{qo} \end{aligned} \tag{A.13}$$

λ_F, λ_D, and λ_Q are not natural rotor variables so far as the stator is concerned. Hence, we transform them into equivalent rotor voltages by means of the following linear transformations [2].

$$\sqrt{3}\, E'_q = \omega L_{MDO}\ell_F^{-1}\lambda_F \tag{A.14}$$

$$\sqrt{3}\, E''_q = \omega L_{MDO}\ell_F^{-1}\lambda_F + \omega L_{MDO}\ell_D^{-1}\lambda_D$$

$$= \sqrt{3}\, E'_q + K_2\omega\lambda_D \tag{A.15}$$

where $K_2 = L_{MDO}\ell_D^{-1}$

$$\sqrt{3}\, E_d'' = -\omega L_{MQO} \ell_2^{-1} \lambda_Q = -L_{AQ} L_Q^{-1} \lambda_Q \tag{A.16}$$

We also define $\sqrt{3}\, V_d = v_d$, $\sqrt{3}\, V_q = v_q$, $\sqrt{3}\, I_d = i_d$, $\sqrt{3}\, I_q = i_q$ and $\sqrt{3}\, E_{FD} = \omega L_{AD} r_F^{-1} v_F$. Substitution of (A.14) - (A.16) in (A.12) and (A.13) together with lengthy simplifications (which are omitted), the following algebraic equations (A.17) for the stator and the differential equations (A.18) for the rotor are obtained [2].

Stator equations

$$\begin{aligned} V_d &= -rI_d - x_q'' I_q + E_d'' \\ V_q &= -rI_q + x_d'' I_d + E_q'' \end{aligned} \tag{A.17}$$

Rotor equations

$$\begin{aligned} \frac{dE_q'}{dt} &= -\frac{1}{\tau_{do}'}\frac{(x_d - x_\ell)}{(x_d' - x_\ell)} E_q' - \frac{1}{\tau_{do}'}\frac{\xi(x_d - x_d')}{(x_d' - x_\ell)} E_q'' \\ &\quad + \frac{x_d'' I_d}{T_3 \xi} + \frac{\xi E_{FD}}{\tau_{do}'} \\ \frac{dE_q''}{dt} &= \frac{1}{T_1} E_q' - \frac{1}{T_2} E_q'' + \left[\frac{1}{T_3} + \frac{1}{T_4}\right] x_d'' I_d + \frac{\xi E_{FD}}{\tau_{do}'} \\ \frac{dE_d''}{dt} &= \frac{-E_d''}{\tau_{qo}''} - \frac{(x_q - x_q'')}{\tau_{qo}''} I_q \end{aligned} \tag{A.18}$$

where $\xi = \dfrac{x_d'' - x_\ell}{x_d' - x_\ell}$, $\dfrac{1}{T_3} = \dfrac{\xi^2}{\tau_{do}'}\dfrac{(x_d - x_d')}{x_d''}$, $\dfrac{1}{T_4} = \dfrac{1}{\tau_{do}''}\dfrac{(x_d' - x_d'')}{x_d''}$

$$\tau_{do}' = \frac{L_F}{r_F}, \quad \frac{1}{T_1} = \frac{1}{\xi \tau_{do}''} - \frac{1}{\tau_{do}'}\frac{(x_d - x_\ell)}{(x_d' - x_\ell)}$$

$$\frac{1}{T_2} = \frac{1}{\tau_{do}''} - \frac{\xi}{\tau_{do}'}\frac{(x_d - x_d')}{(x_d' - x_\ell)}, \quad x_d'' = \omega L_d'', \quad x_q'' = \omega L_q''$$

$$x_d' = \omega L_d', \quad x_q = \omega L_q, \quad x_\ell = \omega \ell_d = \omega \ell_q.$$

(A.17) and (A.18) are the basic equations of the synchronous machine in the hybrid state variables i_d, i_q, E_q', E_q'', E_d''.

Since we have neglected the stator transients, i.e. $\frac{d\lambda_d}{dt}$ and $\frac{d\lambda_q}{dt}$, the differential equations of the stator reduce to algebraic equations.

<u>Electro-mechanical Rotor Equations</u>

To Eq. (A.17) and (A.18), we must add the mechanical equations of motion of the rotor. The rotor angle is chosen relative to a synchronously rotating reference frame moving with constant angular velocity ω_o electrical radians/per sec. If Θ is measured with respect to a fixed reference, then $\theta = \omega_o t + \alpha + \delta$ elec. radians/sec. where α is constant.

The angle $\alpha = \pi/2$ if the d axis leads the q axis and equal to zero if the q axis leads the d axis. We follow the former convention following Ref. [1]. The mechanical equation of motion has been derived in Chapter III as

$$\frac{H}{\pi f_o}\frac{d^2\delta}{dt^2} + D\frac{d\delta}{dt} = P_m - P_g \tag{A.19}$$

i.e. $$M\frac{d^2\delta}{dt^2} + D\frac{d\delta}{dt} = P_m - P_g$$

where P_m and P_g are in p.u. and M has units of sec^2/elec. radian and δ is in radians. As shown in Ref. [1] after the p.u. conversions have been done

$$P_g = \frac{1}{3}(\lambda_d i_q - i_d \lambda_q)\omega \tag{A.20}$$

Using (A.5), (A.6) and (A.14) - (A.16), we can express P_g as

$$P_g = (E''_q I_q + E''_d I_d) + (x''_d - x''_q) I_d I_q \tag{A.21}$$

In terms of the state variables δ and $\omega = \frac{d\delta}{dt}$, (A.19) becomes

$$\frac{d\omega}{dt} = \frac{-D}{M}\omega + \frac{P_m}{M} - \frac{1}{M}[E''_q I_q + E''_d I_d) + (x''_d - x''_q) I_d I_q]$$

$$\frac{d\delta}{dt} = \omega \tag{A.22}$$

Equations (A.17), (A.18), and (A.22) constitute the two axis differential-algebraic model of the synchronous machine and is found adequate for most multimachine simulation studies involving large disturbances. There are five differential equations and two algebraic equations. Several levels of simplification can be made in the model depending on the needed accuracy of results. We consider two such models (i) one axis E'_q model and (ii) the classical model.

A.4 One Axis E'_q Model

In this model there are no damper windings. Consequently $x''_d = x_d$, $x''_q = x'_q = x_q$, $\tau''_{do} = \tau''_{qo} = 0$ and $\xi = 1$. There are no differential equations for E''_d and E''_q. $E''_q = E'_d$ and $E''_d = E'_d$ = constant. The differential equations (A.18) and (A.19) now become

$$\frac{dE'_q}{dt} = -\frac{1}{\tau'_{do}} E'_q + \frac{1}{\tau'_{do}} (x_d - x'_d) I_d + \frac{E_{FD}}{\tau'_{do}} \qquad \text{(A.23)}$$

$$= \frac{1}{\tau'_{do}} (E_{FD} - E)$$

where $E = E'_q - (x_d - x'_d) I_d$.

P_g is derived independently from the expression (A.20) as

$$P_g = E'_q I_q + (x'_d - x'_q) I_d I_q \qquad \text{(A.24)}$$

Hence the rotor equations are

$$\frac{d\omega}{dt} = \frac{-D}{M}\omega + \frac{P_m}{M} - \frac{1}{M}[E'_q I_q + (x'_d - x'_q) I_d I_q] \qquad \text{(A.25)}$$

$$\frac{d\delta}{dt} = \omega$$

The algebraic equations are

$$V_d = -rI_d - x'_q I_q + E'_d$$

$$V_q = -rI_q + x'_d I_d + E'_q \qquad \text{(A.26)}$$

Suppose now we ignore the transient saliency so that $x'_d = x'_q$. Then the algebraic equations (A.26) become

$$V_d = -rI_d - x_d' I_d + E_d'$$
$$V_q = -rI_q + x_d' I_q + E_q' \qquad (A.27)$$

and the differential equations (A.23) and (A.26) become

$$\frac{dE_q'}{dt} = \frac{1}{\tau_{do}'}\left(E_{FD} - E_q' + (x_d' - x_d)I_d\right)$$

$$\frac{d\omega}{dt} = -\frac{D}{M}\omega + \frac{P_m}{M} - \frac{E_q' I_q}{M} \qquad (A.28)$$

$$\frac{d\delta}{dt} = \omega$$

The two algebraic equations (A.27) can be converted to a phasor equation as follows:
From Park's transformation with balanced conditions, the voltage of phase a is [1]

$$V_t = \sqrt{\frac{2}{3}}\,(v_d \cos\theta + v_q \sin\theta) \qquad (A.29)$$

Since $\theta = \omega_o t + \delta + \pi/2$, the phasor representation is

$$\bar{V}_t = V_d e^{j(\delta+\frac{\pi}{2})} + V_q e^{j\delta} = (V_q + jV_d)e^{j\delta} = V_Q + jV_D \qquad (A.30)$$

δ is the angle between the network reference frame and the q axis of the machines (Fig. A.7). The network is assumed to be in steady state. The components of the phasor $\bar{V}_t$ can be expressed either in the network reference frame or the machine reference frame. In the latter case, the q axis is the reference and the voltage phasor is denoted by

$$\tilde{V}_t = V_q + jV_d \qquad (A.31)$$

The algebraic equations (A.27) can then be expressed in either of the two reference frames as

$$\bar{V}_t = -r\bar{I}_t + \bar{E}' - jx_d'\bar{I}_t \qquad \text{(Network Reference)}$$
$$\tilde{V}_t = -r\tilde{I}_t + \tilde{E}' - jx_d'\tilde{I}_t \qquad \text{(Machine Reference)} \qquad (A.32)$$

The components of the network reference frame phasors are the projections along the system D and Q axes and the components of the machine reference frame phasors are along the machine d and q axes. Figure (A.6) shows the equivalent circuit in the network reference frame and Fig. (A.7) shows the components in the two reference frames. The transformation between the two phasors is given by

$$\bar{I}_t = \tilde{I}_t e^{j\delta} \text{ and } \bar{E}' = \tilde{E}' e^{j\delta} \text{ where } \bar{I}_t = I_Q + jI_D,$$

$$\tilde{I}_t = I_q + jI_d, \quad \bar{E}' = E'_Q + jE'_D, \quad \tilde{E}' = \tilde{E}'_q + j\tilde{E}'_d.$$

Notice that δ is the angle between the synchronously rotating network reference frame and the q axis. Q and D are the network reference axes and q and d are the machine reference axes and a phasor can be described in either of these reference frames. The transformation of components between the two reference frames is easily verified as

$$\begin{bmatrix} q \\ d \end{bmatrix} = \begin{bmatrix} \cos\delta & \sin\delta \\ -\sin\delta & \cos\delta \end{bmatrix} \begin{bmatrix} Q \\ D \end{bmatrix};$$

$$\begin{bmatrix} Q \\ D \end{bmatrix} = \begin{bmatrix} \cos\delta & -\sin\delta \\ \sin\delta & \cos\delta \end{bmatrix} \begin{bmatrix} q \\ d \end{bmatrix} \tag{A.33}$$

The transformation applies to both the current and voltage variables.

A.5 Classical Model

To make the final transition to the classical model, we make the assumption that the field voltage E_{FD} may not change very fast so that E'_q remains more or less constant during the 0-1 sec interval of interest in transient stability studies. This is true if τ'_{do} is fairly large. We further assume that r=0 and $E'_q = |\bar{E}'|$ i.e., $E'_d = 0$. This implies that the q axis coincides with E'_q. The phasor diagram now becomes as shown in Fig. A-8. It is seen that $\gamma = \delta$ so that phase angle of $\bar{E}'$ coincides with the rotor angle with respect to the

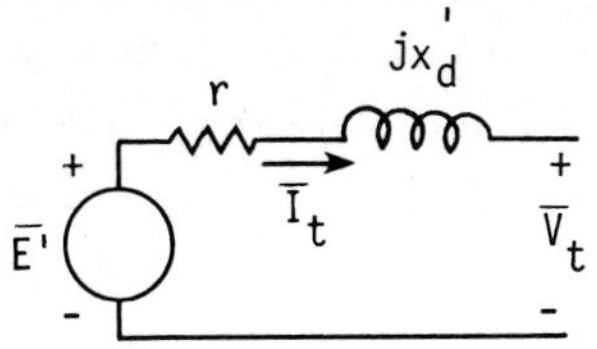

A.6. Voltage behind direct axis transient reactance.

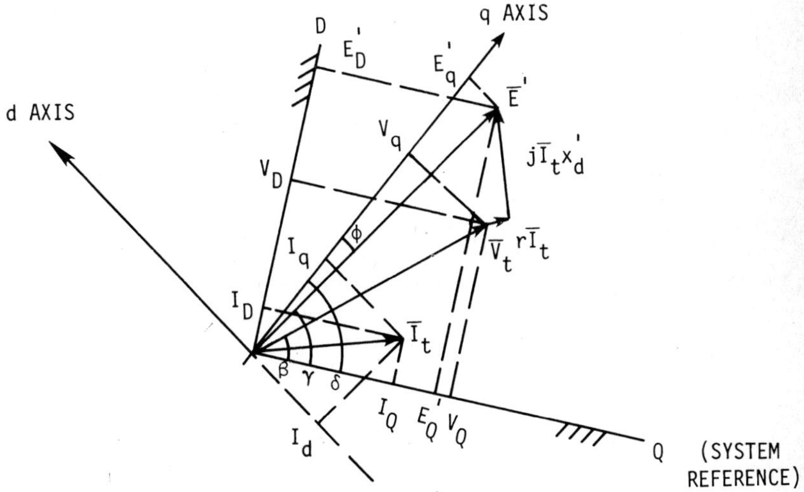

A.7. Phasor diagram for the Eq' one axis model.

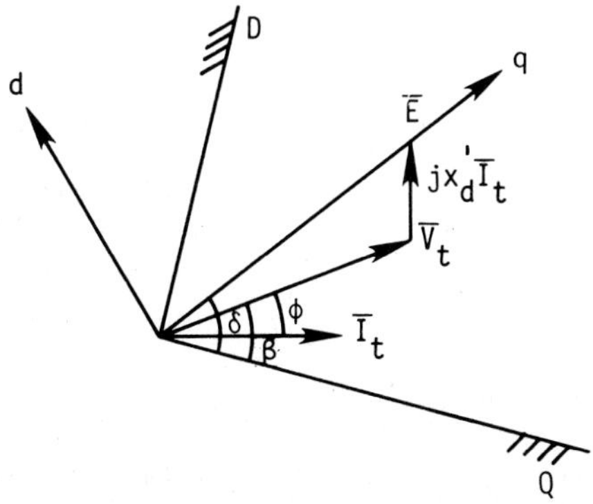

A.8. Phasor diagram for the classical model.

synchronously rotating network reference frame. This obviates the need to transform the phasor quantities from network to machine reference frame and vice versa thus providing a big advantage. Hence we have the oft quoted assumption that the phase angle of the voltage behind transient reactance coincides with the mechanical angle relative to the synchronously rotating reference frame. The expression for electrical power becomes

$$
\begin{aligned}
P_g &= E'_q I_q \\
&= |\bar{E}'||\bar{I}_t|\cos(\delta-\beta+\phi) \\
&= |\bar{E}'||\bar{I}_t|(\cos\delta\cos(\beta-\phi)+\sin\delta\sin(\beta-\phi)) \\
&= E_Q I_Q + E_D I_D \qquad \text{(A.34)} \\
&= \mathrm{Re}[(E_Q+jE_D)(I_Q+jI_D)^*] \\
&= \mathrm{Re}(\bar{E}'\bar{I}_t^*)
\end{aligned}
$$

A.6 Flux Decay Model Used in Chapters III and IV

The model used in Secs. 3.8 and 4.8 is actually a combination of the one axis E'_q model and classical model. The differential equation for E'_q is the one given by (A.23) and the swing equations are

$$
\begin{aligned}
M\frac{d\omega}{dt} &= P_m - P_g \\
&= P_m - \mathrm{Re}[\bar{E}'\bar{I}_t^*] \qquad \text{(A.35)}
\end{aligned}
$$

$$\frac{d\delta}{dt} = \omega$$

The reason for approximating E'_q in the swing equation as in the classical model is to avoid the cumbersome network to machine transformation and vice-versa. Thus while we approximate the electrical power P_g, the flux decay effects are modelled through (A.23).

References

1. Anderson, P. M. and Fouad, A. A., "Power System Control and Stability", (Book), Iowa State University Press, Ames, Iowa, 1977.

2. Riaz, M., "Hybrid-Parameter Models of Synchronous Machines, IEEE Trans., Vol. PAS-93, 1974, pp. 849-858.